QUICK ANSWERS
TO
QUANTITATIVE PROBLEMS
· ·
A Pocket Primer

QUICK ANSWERS TO QUANTITATIVE PROBLEMS
• • • • • • • • • •
A Pocket Primer

G. William Page
Dean, College of Urban and
  Public Affairs
Florida Atlantic University
Ft. Lauderdale, Florida

Carl V. Patton
Vice President for
  Academic Affairs
University of Toledo
Toledo, Ohio

ACADEMIC PRESS, INC.
Harcourt Brace Jovanovich, Publishers

Boston   San Diego   New York
London   Sydney   Tokyo   Toronto

This book is printed on acid-free paper.

Copyright © 1991 by Academic Press, Inc.
All rights reserved.
No part of this publication may be reproduced or
transmitted in any form or by any means, electronic
or mechanical, including photocopy, recording, or
any information storage and retrieval system, without
permission in writing from the publisher.

ACADEMIC PRESS, INC.
1250 Sixth Avenue, San Diego, CA 92101

United Kingdom Edition published by
ACADEMIC PRESS LIMITED
24–28 Oval Road, London NW1 7DX

Cover and Interior Design by Elizabeth E. Tustian

Library of Congress Cataloging-in-Publication Data:

Page, G. William (George William), date.
    Quick answers to quantitative problems : a pocket primer / G.
William Page, Carl V. Patton.
      p.   cm.
    Includes bibliographical references and index.
    ISBN 0-12-543570-3 (alk. paper)
    1. Statistics. I. Patton, Carl V. II. Title.
QA276.12.P33 1991
519.5—dc20   90-46488
                                                                  CIP

Printed in the United States of America
91 92 93 94   9 8 7 6 5 4 3 2 1

CONTENTS

Introduction ix

Acknowledgments xiii

PART 1
How to Describe and Display Data

Chapter 1
Descriptive Statistics 3

Chapter 2
Tabular Analysis 13

Chapter 3
Graphic Techniques 25

PART 2
Basic Ways to Analyze Data

Chapter 4
Scatterplots 43

Chapter 5
Correlation Analysis 49

Chapter 6
Statistical Significance 57

PART 3
Looking at Data Across Time

Chapter 7
Projection Techniques 75
Chapter 8
Annualizing Rates of Change 89
Chapter 9
Shift-Share Analysis 99
Chapter 10
Multiplier Analysis 105

PART 4
How to Obtain and Validate Data

Chapter 11
Sampling 115
Chapter 12
Confidence Levels 133
Chapter 13
Sample Size 151

PART 5
How to Compare Options

Chapter 14
Location Quotient 173
Chapter 15
Indices 179
Chapter 16
Net Benefit Evaluation 189

Glossary 195

Appendices 207
    1 A Checklist for Using Quick Answers to
        Quantitative Problems 208
    2 Words Commonly Confused 210
    3 Perpetual Calendars 212
    4 World Map with Time Zones 214
    5 United States Map with Time Zones and
        Telephone Area Codes 215

Contents   vii

	6	International Distances Chart 216
	7	United States Mileage Chart 218
	8	Political Entities by Size and Population 220
	9	Metropolitan Areas by Size and Population 222
	10	Selected Currencies of the World 225
	11	Greek Alphabet 227
	12	Mathematical Symbols and Operations 228
	13	Review of Algebraic Operations 232
	14	Present Value Tables 233
	15	Distance Conversions 235
	16	Area Comparisons and Conversions 236
	17	Volume Conversions 241
	18	Weight Comparisons and Conversions 243
	19	Temperature Conversions 245
	20	The Use of Scientific Notation..........................246
	21	Prefixes for Metric System Multiples and Submultiples 248
	22	Measurement Unit and Conversion Multipliers.............249
	23	Consumer Price Index 261
	24	Standard Paper Sizes 271
Index		273

Introduction

This primer presents a wide variety of techniques useful in analyzing quantitative data. The techniques are all relatively simple and can be completed quickly with only a pencil and paper in most instances and, in some cases, with basic drawing equipment. These techniques also lend themselves to easy computation using electronic spreadsheets and other common microcomputer software. The techniques are presented with instructions for their use, and examples with contexts and problem solutions are also provided. Although the techniques are not highly complicated, they are extremely useful in a wide variety of public and private sector situations, where conclusions from quantitative data are needed quickly to help make decisions.

Often busy professionals have only limited time to prepare for meetings. Most of the techniques presented here can be completed in a few minutes and many may even be done on the back of an envelope during a meeting. As portable microcomputers of increasing capabilities and decreasing cost are rapidly coming on the market, these quick quantitative techniques will become even more important. Unlike the use of many computer programs, however, these quick techniques are not "black boxes" where the process of quantitative analysis is hidden. Rather, these techniques are open and provide the analyst with important insights into the data and any relationships imbedded in them. The process of using these techniques and their results is easily communicated because they are not complex and are widely used.

These quick methods are also useful as tests to ensure that the results of a complex computer program are reasonable. They can also be used as a first step to decide which of a wide variety of more complex techniques would be most appropriate for a given set of data. There are other situations in which these quick methods can be used to check the reasonableness of someone else's work. Reviewing the work of others often raises questions that the analyst did not address or questions about the appropriateness of the analyst's conclusions. In many circumstances, these quick techniques can be applied to data from a table to explore the potential of a new idea or to gain insights into the conclusions. In summary, these quick quantitative methods can be used as a first approximation to a more complex analysis, as a complete analysis by themselves, or as a check on the reasonableness of a more thorough analysis.

The quick methods presented in this primer are largely statistical in nature. An understanding of statistical theories would be helpful in mastering the methods, but is not essential. This primer does not present statistical theory, nor does it attempt to derive or prove any formulae. The essential components of the methods are presented with clear, step-by-step instructions and examples. Citations to more complete discussions of each technique are provided. This book is small so that it can be carried in a briefcase and be available when the need for quick analysis arises.

No matter the field, today's professionals need to be able to respond quickly to problems and to develop a quick understanding of what the data mean. Whether the person is an aide to a city council member trying to figure out what a citizen opinion poll really means, a private consultant to the health department trying to estimate the number of pregnant teenagers in a particular neighborhood, or the executive director of a small agency who wants to present its budget facts precisely and clearly, the techniques in this book can be helpful.

This primer is divided into five parts. Part 1 presents basic methods for describing and displaying data, including descriptive statistics, tabular analysis, and graphic techniques. Using these methods, the analyst can *describe* and *present* the basic facts in a set of data.

Part 2 includes three chapters about ways to analyze data, including scatterplots, correlation analysis, and statistical significance. With

Introduction

these methods, the analyst can explain the basic relationships among variables, that is, how two or more variables are associated and the extent to which *relationships* between or among variables are or are not due to chance.

Part 3 contains methods for examining data over time, including projection techniques, computing rates of change, analyzing economic change by region, and using multipliers. These methods allow the analyst to *estimate future conditions* based on assumptions about trends and relationships.

Part 4 addresses the question of how to obtain data and assess their validity. Topics include determining the optimum size for a sample, procedures for selecting a sample and obtaining other data, and determining the accuracy of sample estimates. This information allows the analyst to determine *the extent to which the data can be relied upon* as a basis for decision making, that is, to estimate how close values derived from the sample are to values in the population from which the sample was taken.

Part 5 presents several methods that allow the analyst to *compare options*, including the location quotient, indices, and evaluation methods. The location quotient allows us to compare the concentration of a given economic activity between regions, while indices are used to summarize several measurements into a single value that simplifies comparisons among areas, groups, or even countries. The evaluation methods we present focus on economic benefit–cost comparisons, and include the concept of net present value or net benefit as a decision criterion. Each of these methods provides a quantitative way to compare competing options.

While the focus of this book is on analysis, it is essential to remember that good analysis depends on good data and careful data collection techniques as well as on the clear specification of problems, accurate identification of independent and dependent variables, and the application of the proper statistical tests. Quick analysis is not intended to replace other methods, but rather to be used as a first approximation that can be followed by more sophisticated techniques if time permits.

The quick methods presented in this primer are reliable within certain parameters, and any analyst with a knowledge of statistics can easily take the methods to a higher level. For example, data presented in a simple tabular analysis could be further analyzed with inferential statistics. We present a measure of correlation for ordinal data that can also be used to analyze interval data. A more experienced analyst would, however, want to use an interval level correlation measure in this case if time permits.

The tests of significance we present are known as parametric tests, meaning that the statistics make certain assumptions about the parameters that describe the population from which the sample is taken. These assumptions are often violated in practice. For example, the populations are seldom normally distributed, the data may not be interval or ratio scale, samples are seldom simple random samples with replacement, and beyond this, nonsampling errors are seldom considered. While there are non-parametric or distribution-free tests that do not require the knowledge of the precise form or distribution of the population, these tests sometimes require a deeper knowledge of statistics than we assume the reader of this book has, as well as more involved mathematical calculations. When under severe time constraints, we believe that the solution to this dilemma is exactly the one used in practice: relax the assumption of normality, apply parametric tests cautiously, and interpret the results conservatively.

Use the methods in this book to help you find quick answers to quantitative questions, but remember that often you cannot base your conclusions and proposed policies on these statistical tests alone. There must be an underlying logic to the analysis, the conclusions must make sense intellectually, and they must be important as well as statistically significant.

G. William Page
Carl V. Patton

ACKNOWLEDGMENTS

This primer was developed over a number of years during which the authors taught quantitative analysis in several universities. A debt of gratitude is due our colleagues at these and other institutions who reviewed earlier versions of this work, and to our students who suggested improvements over the years. We especially appreciate the critical comments and suggestions received from Curtis Roseman, Michael Romanos, Barry Checkoway, David Forkenbrock, David Lindsley, Jane Patton, John Swift, Catherine Dadlez, and several anonymous reviewers. Elizabeth Tustian and Charles Glaser of Academic Press provided assistance throughout the production of the book.

PART 1
How to Describe and Display Data

Chapter 1
DESCRIPTIVE STATISTICS........

Definition

Descriptive statistics are used to summarize and communicate what we find in quantitative data. We often need simple, quick ways to convey the essential information present in tens or even thousands of individual observations about the subject of interest. We present two types of descriptive statistics: *measures of central tendency* and *measures of dispersion*. Measures of central tendency say something about the "average" characteristic of our subject and are one of the most useful descriptions we can provide. The statistics for this purpose are *mean*, *median*, and *mode*. Measures of dispersion tell us how much the data deviate from the measures of central tendency. They tell us if most of the observations in the distribution (data set) are close in value to the mean or median, or if there is a wide variation in the values. Three common measures of dispersion are the *range*, *variance*, and *standard deviation*.

Mean

The *mean*, or arithmetic average, equals the sum of the values of the observations divided by the number of observations.

$$\overline{X} = \frac{\sum_{i=1}^{n} x_i}{n},$$

where

$\overline{X}$ = *the mean*
Σ = the addition of what follows from the first observation ($i = 1$) to the nth observation,
x = the individual observations (from 1 to n), and
n = the number of observations.

The symbol for the mean is the variable symbol (x in this example) with a bar above it, pronounced: "x bar." This is the symbol for the *sample mean*. If all of the possible observations of the variable, called the population or universe of observations, are collected, then you have the *population mean*. The symbol of the population mean is: μ. This Greek letter is pronounced: "mu" or "mew."

Example 1

The sample data are: 10, 12, 14, 17, 27, 36. $\overline{X} = 116/6 = 19.33$. ∎

Sometimes, one must calculate statistics for data that have been converted from directly measured values into categories. This is usually called grouped data. The formulas used to calculate descriptive statistics for grouped data are modified because one doesn't know where the original measurement belongs within the range of each category of the data available for analysis (Blalock, 1979).

The mean is the most commonly used measure of central tendency. It is particularly valuable because everyone understands it. The mean, however, is sensitive to extreme values. Consequently, it can be a poor measure of central tendency if the data contain a few values that are much larger or smaller than the rest of the data. Data on incomes, where one person with a huge income can distort the average (mean) income statistic, are classic examples of the potential problem.

Example 2

The sample data are: 10, 12, 14, 17, 27, 245. $\overline{X} = 325/6 = 54.167$. Note that this can be a misleading measure of the central tendency of the

Descriptive Statistics

data. See the discussion of the median for use when extreme values are present. ■

Median

The *median* is the measure of central tendency that identifies the mid-most value. The median is like the median strip in a highway: it separates the observations into two equal groups, one lower in value than the median value and one greater in value than the median.

$\tilde{X}$ = the symbol for the median. If x is the variable, it is pronounced "x tilde."

To calculate the median:

1. Order the values from the smallest to the largest;
2. In an odd number of observations, the median is the mid-most value. Ex.: 9, 10, 12, 14, 17, 27, 36. $\tilde{X} = 14$;
3. In an even number of observations, take the average of the two mid-most values. Ex.: 10, 12, 14, 17, 27, 245. $\tilde{X} = 14+17/2 = 15.5$.

Mode

The *mode* is the most frequently occurring value or category of the variable in the data.

The mode is often most easily identified by constructing a *frequency table*, which is an ordering of the data indicating how often each value or category of the data occurs (see Table 1).

Table 1

Frequency Table of Grades on Mid-term Examination

grade (the variable)	frequency (F)
30	2
40	3
50	18
60	26
70	22
80	12
90	6
	$N = 89$

Source: Data developed for example.

The mode or modal value is 60, because more people received the grade of 60 than any other grade. This table presents data in a common frequency table form. Chapter 2, Tabular Analysis, describes the organization of tables to reveal potential relationships.

Range

The *range* is the difference between the smallest and the largest values of the variable in the distribution (data set).

Example 3

There were 221 participants in an adult softball league. The average age of participants was 26.5 years. The youngest was 20 and the oldest was 36. The range of the age of participants is 16 years. ∎

Variance

The *variance* is the arithmetic average of the squared deviations of values from their mean. The symbol for variance is: S^2 for sample data, and σ^2 for population (universe) data. The symbol σ is the lower case letter sigma in the Greek alphabet (see Appendix 11). The formulae for the variance differ if using sample or population data. The following is the formula when we are using data for the population or complete group being studied:

$$\sigma^2 = \Sigma(X - \overline{X})^2/N.$$

When using sample data we must correct for the loss of one degree of freedom by dividing by $n - 1$. See Chapter 6, Statistical Significance, for a discussion of degrees of freedom.

$$S^2 = \Sigma(X - \overline{X})^2/(n - 1).$$

When the data are organized in a frequency distribution (also called a frequency table, see Table 1), the formula must be modified to insure that each observation is included in the calculations. For instance, we must be sure that all 13 softball players aged 22 are included in our calculations. To accomplish this, we modify our formula to multiply

Descriptive Statistics

the squared deviations by the frequency (F) of each value of the variable (X):

$$\sigma^2 = \Sigma F(X - \overline{X})^2/N.$$

When working with data organized in a frequency distribution, we must be sure to multiply the values by the frequency to include all of the observations when we calculate the sample mean or variance.

Example 4

The age distribution of participants in an adult softball league will be used to calculate the variance. We will use Table 2.

Table 2

Frequency Table of the Age of Participants in an Adult Softball League

Age (X)	Frequency (F)	FX	$X - \overline{X}$	$(X - \overline{X})^2$	$F(X - \overline{X})^2$
20	1	20	−6.5	42.25	42.25
21	1	21	−5.5	30.25	30.25
22	13	286	−4.5	20.25	263.25
23	20	460	−3.5	12.25	245.00
24	32	768	−2.5	6.25	200.00
25	33	825	−1.5	2.25	74.25
26	27	702	−0.5	0.25	6.75
27	25	675	0.5	0.25	6.25
28	12	336	1.5	2.25	27.00
29	18	522	2.5	6.25	112.50
30	14	420	3.5	12.25	171.50
31	11	341	4.5	20.25	222.75
32	5	160	5.5	30.25	151.25
33	2	66	6.5	42.25	84.50
34	4	136	7.5	56.25	225.00
35	2	70	8.5	72.25	144.50
36	1	36	9.5	90.25	90.25
TOTALS	221	5844			2097.25

Source: Data developed for example.

Key: FX indicates the product of F (frequency) and X (age) for each observation, $X - \overline{X}$ indicates that we subtract the mean (26.5, the rounded value of the mean) from the value of X, $(X - \overline{X})^2$ indicates that the value of $X - \overline{X}$ is squared, and $F(X - \overline{X})^2$ indicates that we multiply the value of F (frequency) by $(X - \overline{X})^2$.

Since the 221 players are all of the adult softball players (the population or universe), $\sigma^2 = 2097.25/ 221 = 9.49$.

One disadvantage of the variance as a descriptive measure of variation is that the units are difficult to describe. In effect, the units in this example are "years squared." ∎

Standard Deviation

The *standard deviation* is another measure of variation in the data. It has the advantage of being measured in the same units as the variable being analyzed. In this example, the standard deviation is in years of age.

The standard deviation is calculated by taking the square root of the variance:

$\sigma = \sqrt{\sigma^2}$ (using population data), and
$S = \sqrt{S^2}$ (using sample data).

Example 5

$\sigma = \sqrt{\sigma^2} = \sqrt{9.49} = 3.08$.. See the appendix of Chapter 12, Confidence Levels, for further information on calculating the standard deviation of a sample. ∎

The standard deviation tells us that the 221 adult softball players have an "average deviation" of about three years in age from the mean age of 26.5 years. This means that if you picked any baseball player from this group, their average age will be roughly within three years of 26.5. For a more precise explanation of estimating from sample data, see Chapter 12, Confidence Levels. Since the measure is standardized, we can compare the standard deviation in ages among any number of softball leagues.

Coefficient of Variation

The *coefficient of variation* is another measure of dispersion. It is less commonly used than the other measures discussed in this chapter. It is used to compare variation or relative homogeneity in the distributions of data from different groups. When the coefficients of variation for the two sets of data are close in value, then we can assume that there is similar dispersion (variation) of the value being measured in the two

Descriptive Statistics

data sets. The coefficient of variation (CV) is the standard deviation divided by the mean:

$$CV = \frac{S}{\overline{X}}$$

where:

S = the sample standard deviation, and
$\overline{X}$ = the sample mean.

A large coefficient of variation tells us that the data have a lot of variation, i.e., a lot of dispersion about the mean.

Uses of Descriptive Statistics

As the name implies, descriptive statistics are used to describe quantitative data, but they are also used in other statistical procedures. A discussion of the two concepts of scales of measurement and normality is needed to clarify the use of descriptive statistics in other statistical procedures.

Scales of Measurement

There are four *scales of measurement* for quantitative data: *nominal*, *ordinal*, *interval*, and *ratio*. Each is different and the scale of measurement of the data has important implications for the statistical procedures that may be used in its analysis.

Nominal scale data

Nominal scale data clearly distinguish between the observations by using named categories. Examples of nominal scale data includes colors, or sex. Blue is different from red and females are different from males.

Ordinal Scale Data

Ordinal scale data place the observations into ranked order. For example, the observations can be ranked as good, better, or best, or the observations can be ranked on a scale of 1 to 10. Ordinal data not only distinguish between observations but tell us their relative position on a scale.

Interval and Ratio Scale Data

Interval and ratio scale data are measurements in which the units of measurement can be continually subdivided for greater accuracy. Observations of distance, time, and weight are examples of common interval scale measurements. Ratio scale data have an absolute or non-arbitrary zero that allows ratio comparisons, enabling us to say that some distance is twice as great as another. For example, 100 pounds are twice as much as 50 pounds. Because the Fahrenheit temperature scale has an arbitrary zero, we cannot say that 100 degrees F is twice as hot as 50 degrees F.

Normality

Normality refers to the distribution of the data. If a simple frequency polygon or frequency distribution of the data has a classic bell-shaped curve, then the data are normally distributed. In practice, many sets of data are sufficiently close to the classic bell-shape to be considered "normally distributed." This is because many phenomena have most observations in the middle of the range of observations with a small number of high values and a small number of low values. Refer to Chapter 12, Confidence Levels, for a more complete discussion of the normal curve.

In the normal distribution, the mean, mode, and median are all the same value and are located in the center of the distribution. Therefore, one quick test to determine whether the data are normally distributed is to compare the mean, mode, and median to see if they are reasonably close in value. Since almost all observations are within three standard deviations of the mean in the normal distribution, another test is to confirm that there are few extreme values outside three standard deviations. If the graph of the data looks bell-shaped and it passes these two quick tests, you should feel confident in assuming the data are normally distributed.

Inferential Statistics

There are two categories of *inferential statistics:* parametric and nonparametric. *Parametric statistical procedures* are powerful procedures in helping you analyze data and reach conclusions. Parametric statistical procedures require interval or ratio scale data and that the data be normally distributed. Most parametric statistical procedures are suffi-

Descriptive Statistics

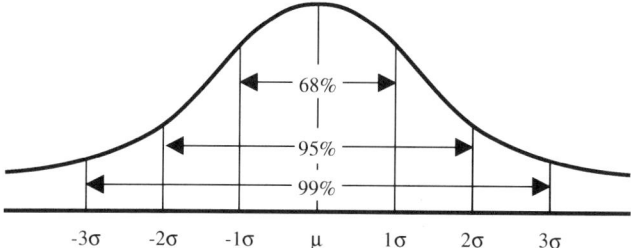

Figure 1. Normal Distribution.

ciently robust so that we can use them with some confidence when our data meet the simple test of normality described above.

If the data are nominal or ordinal scale or if the data are so *skewed* (i.e., have a distribution that deviates from a normal bell-shaped curve) that they cannot be considered normally distributed, then there are *nonparametric statistical procedures* available. Nonparametric statistical procedures are able to perform a wide range of statistical analyses that are comparable to those that could be accomplished using parametric statistical procedures if the data fit the assumptions required by parametric procedures. There are also several other distributions, for example, the binomial or the Poisson distributions, for which powerful statistical procedures have been developed. The disadvantage of nonparametric statistical procedures is that they are not as powerful as parametric statistical procedures. Because of the limitations of the data, these procedures are not able to discriminate as well or allow us to reach conclusions as readily as parametric statistical procedures.

In performing quick analyses, descriptive statistics are essential. Descriptive statistics provide us tools that help us understand and summarize data. Descriptive statistics also provide us a vocabulary that we can use to communicate information concerning quantitative data.

Further Readings

Andrews, F. M., L. Klem, T. Davidson, P. O'Malley, and W. Rodgers, *A Guide for Selecting Statistical Techniques for Analyzing Social Science Data, Second Edition.* Ann Arbor: Survey Research Center, Institute for Social Research, The University of Michigan, 1981.

Blalock, Hubert M., Jr., *Social Statistics, Revised Second Edition.* New York: McGraw-Hill Book Company, 1979.

Horwitz, Lucy, and Lou Ferleger, *Statistics for Social Change.* Boston: South End Press, 1980.

Krueckeberg, Donald A., and Arthur L. Silvers, *Urban Planning Analysis: Methods and Models.* New York: John Wiley and Sons, 1974.

Matlack, William F., *Statistics for Public Policy and Management.* Boston: Duxbury Press, 1980.

Smith, G., *Statistical Reasoning.* Boston: Allyn and Bacon, 1985.

Tufte, Edward R., *Envisioning Information.* Cheshire, Connecticut: Graphics Press, 1990.

Chapter 2
TABULAR ANALYSIS

Definition

Although a great amount of data can be summarized as means, medians, modes, percentages, and proportions, there is often a need to summarize data in tabular form in order to allow the reader to examine relationships among components of the data. *Tabular analysis* involves displaying data in a logical, consistent format that permits easy and accurate interpretation.

Method

Since analysts typically examine and display ordinal or interval data that are usually positive in value, tables should be visualized as being contained in the x positive, y positive quadrant of the coordinate system. The values of variables should be organized from low to high along the x (horizontal) and y (vertical) axes. This results in a layout in which data that are positively correlated will tend to fall on the diagonal that runs from the lower left to the upper right corner of the table. A negative correlation will be described by data that tend to lie on the opposite diagonal, the one running from the upper left corner to the lower right.

The relationships are most obvious in square tables but can also be visualized with a little practice in tables that are not square.

Please note that the patterns that indicate positive and negative relationships are valid *only* when the variables have been ordered from low to high in *both* directions. Figure 1 in Chapter 4, Scatterplots, presents a negative relationship for individual data points which are depicted in a "scatterplot," these points being contained in the x positive, y positive quadrant of the coordinate system.

When constructing tables, follow these basic rules:

- Specify the independent and dependent variables. (Recall that the independent variable is the one we suspect affects the behavior of the other, or dependent, variable.)
- Divide the data into categories of the independent variable.
- Describe each of these categories in terms of the dependent variable values.
- Be sure the categories are mutually exclusive and exhaustive.
- Use evenly divided categories whenever possible.
- Report missing values and responses not analyzed because they were not applicable.
- Discuss the categories of the independent variable in terms of the dependent variable.
- Identify the data sources.

We also suggest that you follow these conventions for tabular analysis when they help clarify the presentation:

- Make the independent variable the rows.
- Make the dependent variable the columns.
- Calculate percentages across the rows to equal 100%, and read down the columns when interpreting the data.
- Give the table a title that is both descriptive and easily remembered.
- Round off cell percentages to whole numbers.

- Report only row totals (and/or column totals), as individual cell values can be recomputed if necessary.

In performing tabular analysis, we are attempting to specify the more important variables in order to illustrate how they are related and to draw out the essence of the data. Before constructing a table, we must understand the nature of the data, determine the independent and dependent variables, and decide how to categorize the data. After laying out the initial table, we may need to reconfigure it in order to present better the meaning of the data.

Example 1

A sample survey of residents in the census tracts surrounding the local ball park was conducted. The respondents were surveyed about their attitudes toward night baseball games being permitted at the park. At present, only daytime games are permitted.

The data were initially presented publicly in the form given in Table 1, showing that the respondents were evenly split over the issue of whether to permit night baseball, with some difference between the attitudes of men and women. The basic data are in the appendix to this chapter.

Table 1

Residents Are Evenly Split on Night Baseball

	Approve Night Baseball		
	No	Yes	Total
Females	54%	46%	100% (240)
Males	44%	56%	100% (180)
Total	50% (210)	50% (210)	100% (420)

Missing Observations = 12

Source: Sample survey of residents

Following the guidelines presented earlier, the data were organized properly. Note that:

1. The data are divided into independent and dependent variables (gender of respondent and attitude about night baseball; that is, gender affects one's attitude about night baseball).
2. The independent variable is divided into mutually exclusive and exhaustive categories (males/females).
3. Each of these categories is described in terms of the dependent variable values (no/yes regarding approval of night baseball), and is also mutually exclusive and exhaustive (if some people had been undecided we would have added a third column or shown this in a footnote).
4. The categories of the variables are evenly divided (in this example there was no alternative).
5. The twelve people who did not respond to the survey are shown as missing observations.

We would then report on the categories of the independent variable in terms of the dependent variable (since we calculated percentages across we read down):

"Fifty-six percent of men approve of night baseball compared with 46% of women."

This basic observation should suggest to us that there may be more to the data than appear in the table, and that at least the title to the table may be a little misleading. We can explore the data in more depth through multivariate tabular analysis. ∎

The above example is known as a *bivariate* tabular analysis since it involves only two variables (gender of respondent and attitude about night baseball). *Multivariate* tables are also useful. These tables have more than one independent variable that is used to explain the variation in the dependent variable. In the above example, we may have a hunch that something other than one's gender affects attitudes toward night baseball. This might be homeowner status, age, or a similar characteristic. (The difference might also be an error as a result of the way we selected our sample, but let us assume that this is not the case and that the sample was selected correctly.)

Example 2

Assume that we thought age has an impact on one's views toward approval of night baseball. The previously given data could be reconfigured to present attitudes by both gender and age, assuming we had collected information about the age of respondents. Since age of respondent was asked in the survey, we were able to prepare the multivariate description shown in Table 2 by using the data appended to this chapter.

Table 2

Who Approves of Night Baseball?

	Approve	
	No	Yes
Females		
40 and Older	90	30
Under 40	40	80
Males		
40 and Older	60	40
Under 40	20	60

Missing Observations = 12

Source: Sample survey of residents.

For ease of interpretation and presentation, Table 3 was generated by converting these data to percentages.

The data displayed in the table in this way suggest several conclusions:

1. Age does apparently affect one's attitude about night baseball.
2. Among both men and women, greater percentages of younger people approve of night baseball.
3. No matter the age category, a greater percentage of men than women favor approving night baseball.
4. Age has a stronger effect on attitudes toward night baseball than does one's gender.

Table 3

Younger People Support Night Baseball

	Approve of Night Baseball		
	No	Yes	Total
Females			
40 and Older	75%	25%	100% (120)
Under 40	33%	67%	100% (120)
Total Females	*54%*	*46%*	*100% (240)*
Males			
40 and Older	60%	40%	100% (100)
Under 40	25%	75%	100% (80)
Total Males	*44%*	*56%*	*100% (180)*
Total			
40 and Older	68%	32%	100% (220)
Under 40	30%	70%	100% (200)

Missing Observations = 12

Source: Sample survey of residents.

As we mentioned above, home ownership might also have an impact on one's attitude toward night baseball, under the assumption that home owners fear that night baseball would negatively affect their neighborhood and reduce their property values. Therefore, since data had been collected about home ownership, the above analysis should be recomputed, using home ownership as the independent variable. We have done this in Table 4.

Table 4 suggests that home ownership status affects the views of females toward night baseball. Males, even among those who are home owners, still tend to support night baseball. To this point, we have found that, overall, residents are split 50/50 on the issue of whether to permit night baseball; that younger people tend to favor night baseball; that men, no matter the age tend to favor night baseball; and that home ownership affects negatively the view of females toward night baseball.

Table 4

Female Home Owners Oppose Night Baseball

	Approve of Night Baseball		
	No	Yes	Total
Owners			
Females	59%	41%	100% (135)
Males	44%	56%	100% (90)
Non-owners			
Females	48%	52%	100% (105)
Males	44%	56%	100% (90)
Total			
Owners	53%	47%	100% (225)
Non-owners	46%	54%	100% (195)
Grand Total	50%	50%	100% (420)

Missing Observations = 12

Source: Sample survey of residents.

We have now determined that there was more to the data than first publicly reported, and so we construct Table 5 (a multivariate table), to show the relationships that we have identified in the data. Because age seems to be the most important independent variable, we use that as the major division among the data, followed by gender and home ownership. We leave it to you to draw conclusions from the data and to devise a title for the table.

Various policy implications might be drawn from these data. The purpose here, however, was to illustrate how to construct tables in a way that identifies relationships among variables and yields maximum information. For this small set of data, the relationships we uncovered may have been readily apparent in the raw data. For a larger set of less obvious data, the process described here can be useful for revealing relationships among variables. ∎

Table 5

Your Title Goes Here

Ages	Approve of Night Baseball		
	No	Yes	Total
40+			
Females			
Owners	77%	23%	100% (65)
Non-owners	73%	27%	100% (55)
Males			
Owners	60%	40%	100% (50)
Non-owners	60%	40%	100% (50)
Total 40+	*68%*	*32%*	*100% (220)*
<40			
Females			
Owners	43%	57%	100% (70)
Non-owners	20%	80%	100% (50)
Males			
Owners	25%	75%	100% (40)
Non-owners	25%	75%	100% (40)
Total <40	*30%*	*70%*	*100% (200)*
Grand Total	50%	50%	100% (420)

Missing Observations = 12

Source: Sample survey of residents.

A variety of other types of tables are useful for displaying data. Several of these follow as examples of ways to summarize population change over time, age distribution by gender, and income distribution by age of head of household. Sample layouts are shown in Tables 6, 7, and 8, drawn from a study conducted by one of the authors.

Example 3

Table 6 presents population change over time for numerous municipalities. Both the number of persons and percentage change for several time periods are shown. This allows the reader to compare both actual numbers of persons and relative change among municipalities. Note that percentage change is given for the entire 15-year period, as well as the annual percentage change for three different time periods. (Methods for annualizing rates of change are given in Chapter 8.) ∎

Table 6

Population Change from 1970 to 1985 for Selected Jurisdictions

				Percentage Change			
	Population			Annual		Total	
Area	1970	1980	1985	1970–80	1980–85	1970–85	1970–85
Indian Hill	5,651	5,521	5,191	−0.23	−1.23	−0.56	−8.14
Montgomery	5,683	10,088	11,356	5.91	2.40	4.72	99.82
Blue Ash	8,324	9,506	11,000	1.34	2.96	1.88	32.15
Sharonville	10,985	10,108	10,708	−0.83	1.16	−0.17	−2.52
Evendale	1,967	1,954	2,118	−0.07	1.63	0.49	7.68
Madeira	6,713	9,341^a	8,745	3.36	−1.31	1.78	30.27
Terrace Park	2,266	2,044	1,982	−1.03	−0.61	−0.89	−12.53
Mariemont	4,204	3,295	3,012	−2.41	−1.78	−2.20	−28.35
Newtown	2,047	1,817	1,981	−1.18	1.74	−0.22	−3.22
Sycamore Twp.	22,733	20,758	21,687*	−0.90	0.88	−0.31	−4.60
Symmes Twp.	3,726	5,861	6,646*	4.63	2.55	3.93	78.37
Anderson Twp.	25,887	34,504	35,448*	2.92	0.54	2.12	36.93
Hamilton County	924,017	873,224	859,651	−0.56	−0.31	−0.48	−6.97

Sources: 1970 and 1980, U.S. Census of Population and Housing for Ohio; 1985, Hamilton County Regional Planning Commission data files.

*July 1984; the other population figures are for the beginning of the year indicated.

^{a}Most of this growth was due to annexation; 1,225 housing units were added through annexation between 1970 and 1979.

Note: Both numerical and percentage change values are shown for each geographic area. Percentage change is shown for both the total fifteen year period and the three sub-periods. The percentage change for the three sub-periods is an annualized figure, that is, is adjusted to reflect the effects of year-to-year compounding.

Example 4

Table 7 displays income data for a selected part of the population of a given community. Both the number and percentage of households with various incomes are shown by age of the head of household. This permits us to see the actual number of households in the various age categories. For the three age categories, the mean and median household incomes are also given. ∎

Table 7

Income Distribution in 1979 (1980 Census)
for the Older Population in Indian Hill

	Age of Head of Household					
	55–59		60–64		65+	
Household Income	No.	%	No.	%	No.	%
$75,000+	126	53.8	56	33.9	54	23.2
50,000–74,999	38	16.2	50	30.3	7	3.0
40,000–49,999	20	8.5	14	8.5	22	9.4
30,000–39,999	13	5.6	12	7.3	14	6.0
20,000–29,999	26	11.1	17	10.3	56	24.0
10,000–19,999	0	0.0	5	3.0	57	24.5
Less than 10,000	11	4.7	11	6.7	23	9.9
Total	234	100.0	165	100.0	233	100.0
Mean	$95,915		$81,537		$47,054	
Median	$75,001		$61,750		$26,298	

Source: 1980 U.S. Census of Population and Housing, Summary Tape File 3.

Note: The income categories are mutually exclusive and exhaustive and the summary statistics on income are provided for each age group. Some columns may not equal 100% because of rounding.

Example 5

Table 8 presents detailed data on the age and gender composition of the population for three time periods. Again, both actual numbers and percentages are given. This table allows us to examine change over time by gender. In addition to indicating the source of data, a note is given to avoid confusion about minor data irregularities. ■

Table 8

Indian Hill Population Distribution 1970–1980–1985

Age Group	Males						Females					
	Number			Percentage			Number			Percentage		
	1970	1980	1985	1970	1980	1985	1970	1980	1985	1970	1980	1985
75+	58	86	114	2.0	3.1	4.3	82	118	129	2.9	4.3	4.9
65–74	186	210	265	6.5	7.6	10.0	170	213	250	6.1	7.7	9.5
55–64	328	435	481	11.5	15.8	18.2	319	403	478	11.4	14.6	18.2
45–54	494	478	385	17.3	17.3	14.6	508	510	452	18.1	18.5	17.2
35–44	358	336	199	12.6	12.2	7.5	447	421	282	15.9	15.2	10.7
25–34	121	162	238	4.2	5.9	9.0	149	187	209	5.3	6.8	8.0
15–24	492	496	584	17.3	18.0	22.1	434	426	535	15.5	15.4	20.3
5–14	695	438	259	24.4	15.9	9.8	601	404	209	21.4	14.6	7.9
0–4	117	116	120	4.1	4.2	4.5	99	82	87	3.5	3.0	3.3
Total	2,849	2,757	2,645	100.0	100.0	100.0	2,809	2,764	2,631	100.0	100.0	100.0

Sources: 1970, U.S. Bureau of the Census, Table P.1; 1980, U.S. Census of Population and Housing, Summary Tape File 2; 1985, LOGOS Planning and Research Associates, Cincinnati, Ohio.

Note: Summary Tape File 2 contains 100% count data. Some published data for Indian Hill have been taken from Tape File 4, which contains sample data. Thus the two sets of data may contain minor discrepancies, but these do not significantly affect the analysis. The unequal age categories result from the way in which the original data were summarized.

Appendix

Base data for the examples:

Gender	Age	Owner	Approve Night Ball
Female	<40	yes	yes = 40
Female	<40	yes	no = 30
Female	<40	no	yes = 40
Female	<40	no	no = 10
Female	40+	yes	yes = 15
Female	40+	yes	no = 50
Female	40+	no	yes = 15
Female	40+	no	no = 40
Male	<40	yes	yes = 30
Male	<40	yes	no = 10
Male	<40	no	yes = 30
Male	<40	no	no = 10
Male	40+	yes	yes = 20
Male	40+	yes	no = 30
Male	40+	no	yes = 20
Male	40+	no	no = 30

Source: Sample survey of residents.

Further Readings

Babbie, Earl R., *Survey Research Methods*. Belmont, California: Wadsworth, 1973.

Patton, Carl V., and David S. Sawicki, *Basic Methods of Policy Analysis and Planning*. Englewood Cliffs, New Jersey: Prentice-Hall, Inc., 1986.

Tukey, John W., *Exploratory Data Analysis*. Reading, Massachusetts: Addison-Wesley Publishing Company, Inc., 1977.

Chapter 3
GRAPHIC TECHNIQUES ......

Definition

Graphic techniques are diagrams, drawings, and pictographs that can be used to present data quickly and effectively. They can be used to summarize information, portray relationships, show change over time, and present conclusions. Graphic techniques include charts, time series plots, maps, diagrams, and even cartoons.

Method

Analysts use various types of graphic methods to present data visually. While tabular analysis is a graphic technique, this chapter focuses on methods that are more pictorial. As is the case with tabular analysis, the principal purpose of these methods is to summarize and simplify complex relationships. Graphic presentation methods should be selected for their simplicity, clarity and usefulness for conveying the meaning of a set of data to the intended audience.

We expand upon a number of principles by Patton and Sawicki (1986, p. 84) that should be kept in mind when preparing graphic displays:

- Give all graphics a title.
- Divide the data into mutually exclusive and exhaustive categories.
- Use evenly divided categories whenever possible.
- Round off final data to whole numbers and percentages when this would not be misleading.
- Time should run from left to right and from bottom to top.
- Magnitudes should run from left to right and from bottom to top.
- Label trend lines directly, if possible, in order to reduce the amount of information that must go into a legend.
- Maps should include legends that identify the variables being presented and their respective values. Orient maps so that north is at the top. Include a north arrow and scale.
- Shading or tones should be selected that will photocopy and telecopy clearly.
- Beginners should avoid using three-dimensional graphics which tend to distort the relationships among magnitudes. More experienced analysts may wish to consult Tufte (1990).
- Cite your sources.

Pie Charts

Pie charts illustrate proportions or shares of the whole. Their categories are computed as proportions of the 360 degree circle. If comparisons are made between or among pie charts, each chart must be the same size in order to permit accurate comparisons of areas.

Example 1

Figure 1 shows agency revenue as a proportion of the 360 degree circle. For example, the state revenue portion of 20% is a 72 degree wedge (360 degrees × .20 = 72 degrees). ■

Graphic Techniques

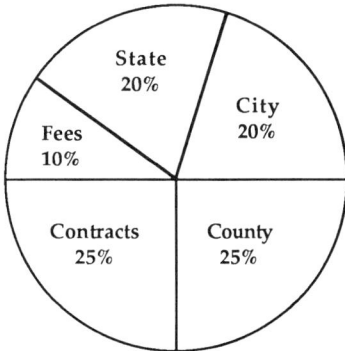

Figure 1. Agency Revenue by Source for Fiscal Year 1991–92. *Source*: Original Data for Our Town.

Bar Charts

Bar charts compare the differences among mutually exclusive categories of non-continuous nominal or ordinal data. The heights of the bars depict the frequencies (e.g., volume, number, or amount) of the categories of the variable being studied. The bars are drawn so that they do not touch because the data being displayed are non-continuous.

Example 2

Figure 2 shows the number of persons in each racial category in Our Town. The bars are separated because race is a *non-continuous* variable, that is, race represents separate categories of data, whereas income data, for example, are *continuous* data that are grouped into convenient categories. Both axes are labeled with the respective variable name and value categories specified (e.g., *population*, 2500, and *race*, white). ■

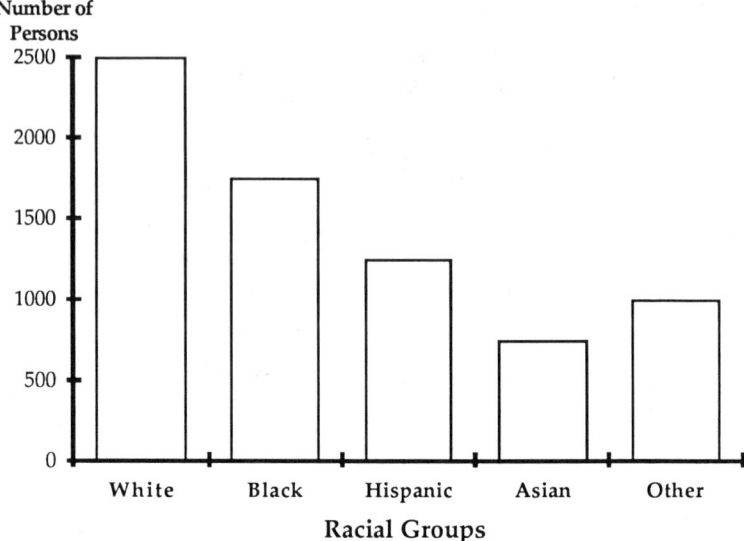

Figure 2. Racial Composition of Our Town. *Source*: Original Data for Our Town, 1991.

Histograms

Histograms describe differences among categories of continuous interval or ratio data. In practice, however, the distinction between bar charts and histograms is not consistently made. (For example, histograms are sometimes used to depict a discrete distribution such as the sum of n rolls of a die and how it can be approximated by a normal distribution.)

Example 3

Figure 3 shows the number of households in seven mutually exclusive income categories. The bars are drawn touching one another because income is a continuous variable that has been grouped into seven categories for graphic purposes. Income could have been grouped into any number of categories depending on the range and distribution of data and the purpose of the analysis. ∎

Graphic Techniques

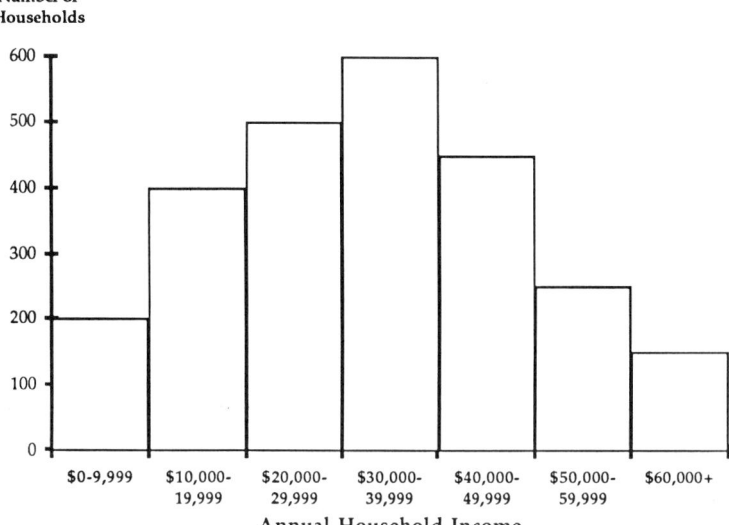

Figure 3. Household Income Distribution in Our Town. *Source*: Original Data for Our Town, 1991.

Dot Diagrams

Dot diagrams are used when the variables have many categories or the data are grouped more finely than for histograms. Scatterplots, covered in Chapter 4, are an extension of this concept.

Example 4

Figure 4 shows the number of housing units at each $10,000 level of appraised value. For example, there are 45 units appraised at approximately $100,000. To create a dot diagram we format the data into convenient categories by either rounding the values to the nearest category value (e.g., $1,000, $5,000 or $10,000) or group the data into categories (e.g., categories of $10,000—appraised values falling between $90,000 and $99,999 or between $95,000 and $104,999). The selection of categories $10,000 in width was arbitrary; $5,000 or even $1,000 categories could have been used, but with the risk that the

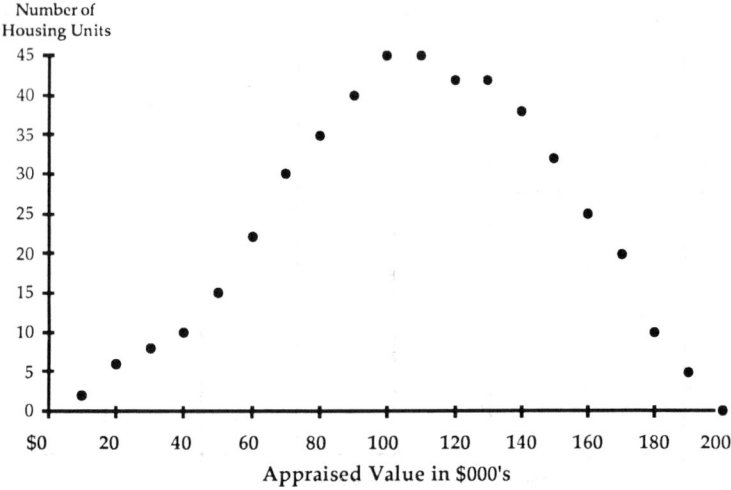

Figure 4. Value of Housing Units in Census Tract 315. *Source*: Special Census for Our Town, 1991.

diagram would have become more difficult to understand. Sometimes the dots are connected to produce a *frequency polygon*. Dot diagrams, frequency polygons or line graphs are appropriate when each x value has only one y value. When the data have not been grouped or summarized and therefore there can be more than one y value for each x value, the data should be displayed as a scatter diagram (see Chapter 4, Scatterplots). ∎

Time Series Diagrams

Time series diagrams show change over time for a variable. Beginning and ending data points alone can be plotted if the data describe relatively constant change. Intermediate values should also be plotted if they reveal important changes that would be hidden if only the beginning and ending points were plotted. Show as few data points as needed to present the story accurately and clearly. More than one variable can be plotted on a diagram, but too many variables or data points will defeat the purpose of the graphic.

Graphic Techniques

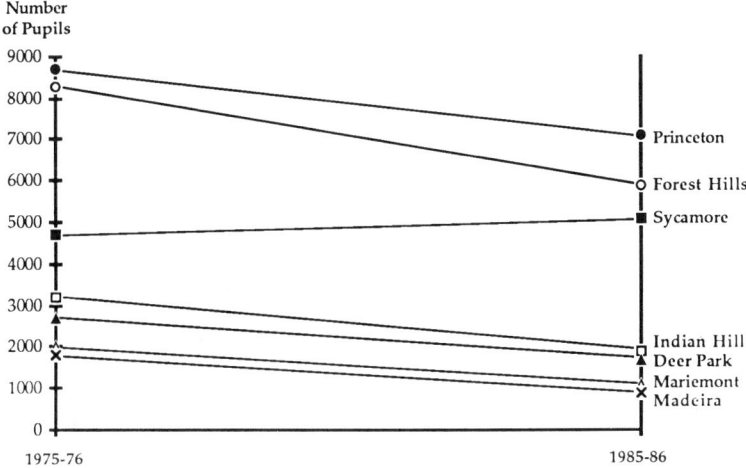

Figure 5. Change in Public School District Enrollment 1975–76 to 1985–86. *Source*: Hamilton County, Ohio Regional Planning Commission annual school enrollment reports.

Example 5

In Figure 5, public school enrollment trends are plotted for seven school districts from a study conducted by one of the authors. This diagram shows actual number of students. To simplify the presentation, only the beginning and ending years are plotted. In this case the changes were relatively constant from year to year. If there had been year-to-year fluctuations, a diagram such as Figure 7, which shows the intermediate points, would have been more appropriate. ∎

Example 6

In Figure 6, the data in Figure 5 are replotted as percentages to show the change over time. For ease of reference, the dashed line identifies the school district that is the focus of the study. ∎

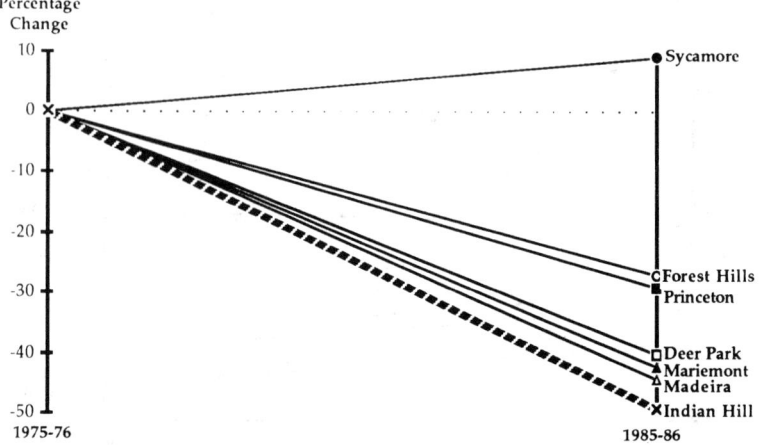

Figure 6. Percentage Change in Public School District Enrollment 1975–76 to 1985–86. *Source*: Hamilton County, Ohio Regional Planning Commission annual school enrollment reports.

Example 7

Figure 7 is used to compare change by decade among eight census tracts. Unlike Figure 5, intermediate values are plotted as well as the beginning and ending points, because presenting the geometric change between 1950 and 1990 as straight lines would be misleading. ■

Example 8

Figure 8 shows change over time for both public and private school enrollment in Indian Hill, as well as total change. More than two categories could be used, but when more than four categories are used the comparisons become difficult to understand. ■

Graphic Techniques 33

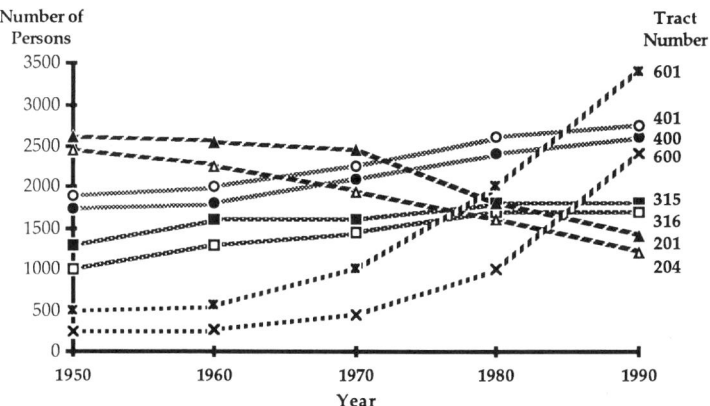

Figure 7. Population Change in Selected Census Tracts in Our Town 1950 to 1990. *Source*: U.S. Bureau of the Census data for Our Town.

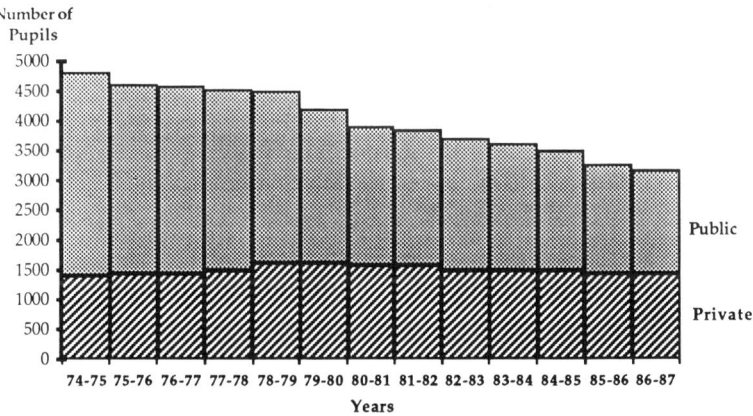

Figure 8. Public and Private School Enrollment in Indian Hill 1974–75 to 1986–87. *Source*: Hamilton County, Ohio Regional Planning Commission annual school enrollment reports.

Population Pyramids

These graphics are used to show population distribution by sex and age. They can also be used to compare proportions of populations for various time periods. Either actual numbers of persons by age and sex can be displayed or the data can be converted into percentages to permit comparison among population groups that vary in size.

Example 9

The diagrams in Figure 9, from an analysis by one of the authors to estimate inter-censal population, compare the population distributions by age and sex for three time periods. The data for Figure 9 were presented in Table 8 in Chapter 2, Tabular Analysis. These population pyramids are merely back-to-back relative frequency histograms turned on their sides. That is, the left side of the population pyramid depicts the relative numbers or percentages of males in various age categories and the right side of the population pyramid depicts the same relationship for females. Here we are comparing percentage distributions. We could also compare actual numbers of persons because the data are for one geographic area. The use of percentages, however, allows us to compare the composition of populations that differ in size. Because of the way the data for 1970 were reported, the first age category contains only five years, where the other categories contain ten years. The oldest age group must be open-ended so as to include all older age groups. ∎

Example 10

In Figure 10, we overlay the 1970 and 1985 data from Example 9 to highlight differences. Again percentages are compared, but actual numbers of persons could also be compared. This type of diagram allows us to see the proportionate change over time in each age group of the population. ∎

Graphic Techniques

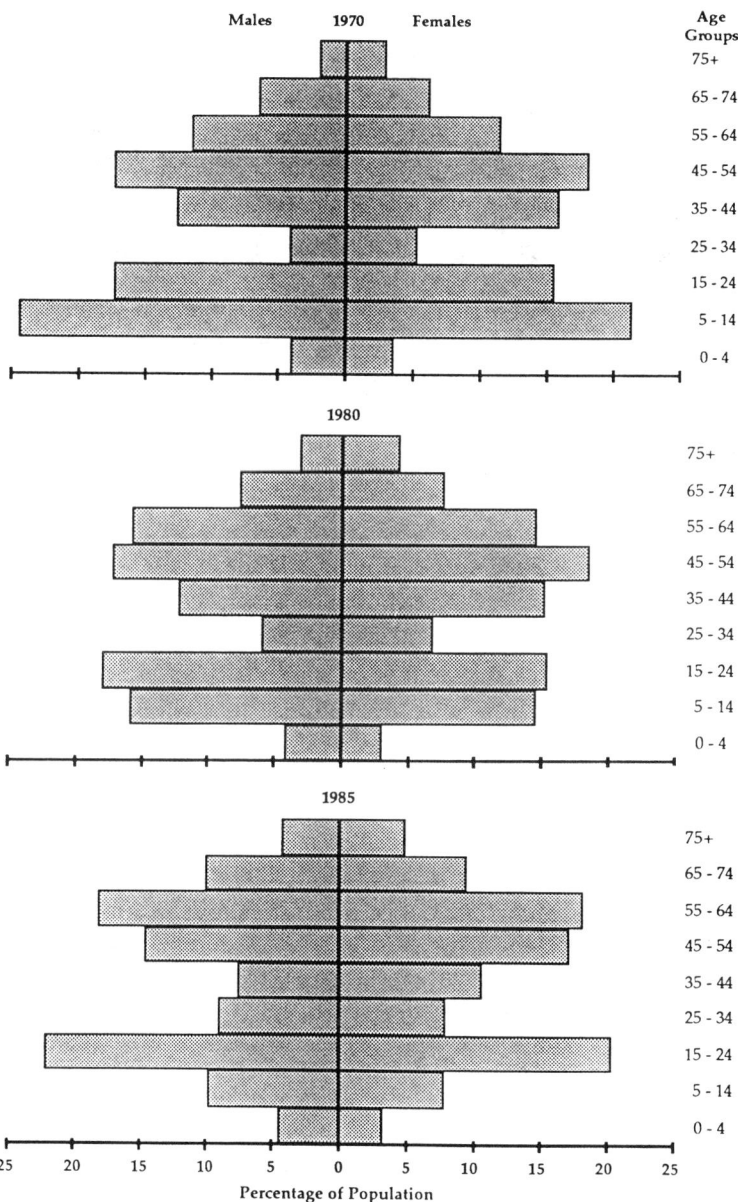

Figure 9. Population Distribution of Indian Hill in 1970, 1980, and 1985. *Source*: U.S. Bureau of the Census, Summary Tape 2.

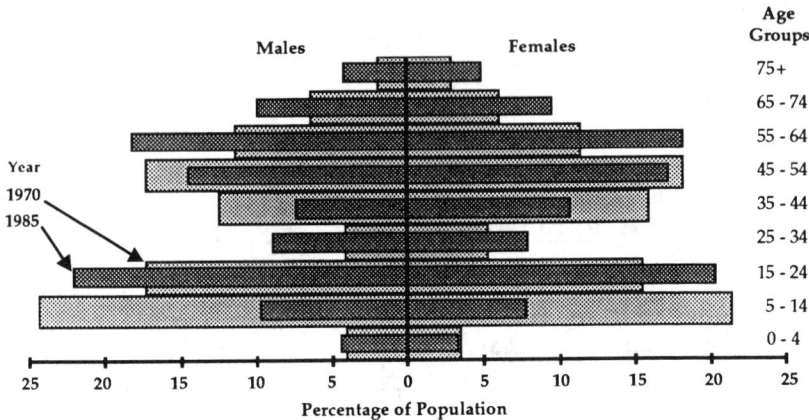

Figure 10. Comparison of 1970 and 1985 Population Distributions of Indian Hill. *Source*: U.S. Bureau of the Census, Summary Tape File 2.

Maps

When spatial relationships need to be displayed, *maps* can be used. For thematic mapping, a gradation of tones can be used to show differences in values of a variable. Symbols can be used to indicate landmarks and key facilities. A series of transparent overlays can be used to show combinations of variables.

Example 11

In Figure 11, by using a gradation of tones, variation in dwelling unit densities are shown for an urban area. Of course, symbols could be added to show the location of public facilities and landmarks. The more data that are added, however, the more difficult it becomes to recognize patterns. Be sure to include a legend, scale, and north arrow and to cite your sources. Many inexpensive computer mapping programs are readily available. ■

Graphic Techniques

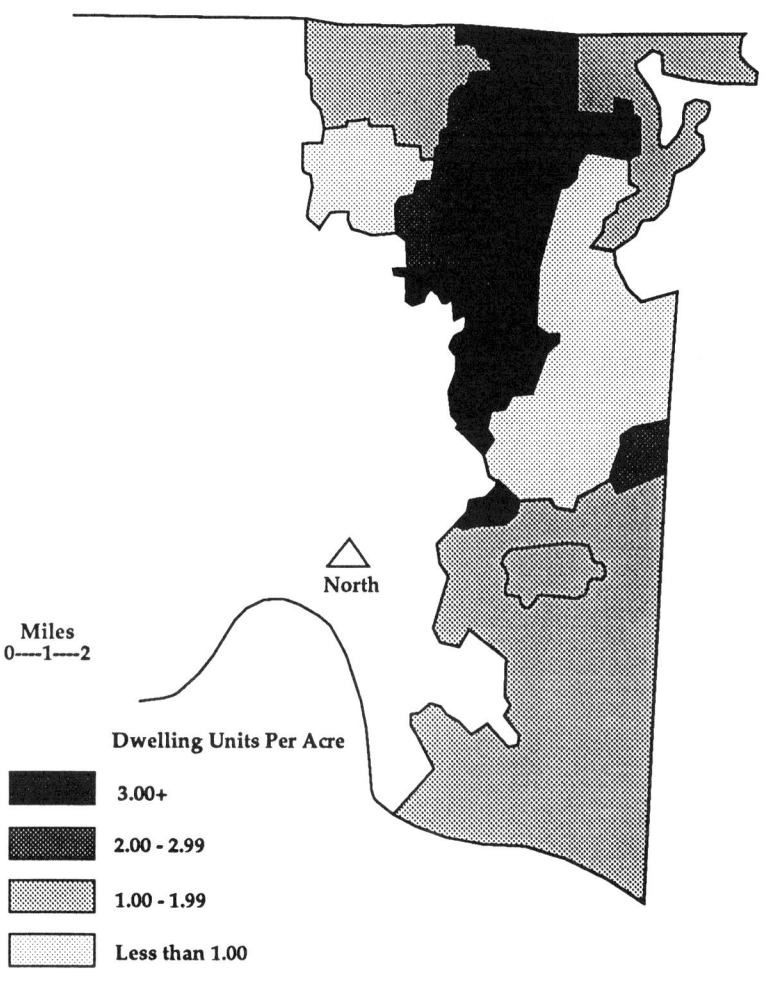

Figure 11. Densities in Selected Jurisdictions, Hamilton County, Ohio. *Source*: Hamilton County, Ohio Department of the Building Commissioner, 1988.

Process Diagrams

Diagrams and flow charts can be used to show organizational structures and decision processes, to compare alternatives, and to portray other components of planning and management activities.

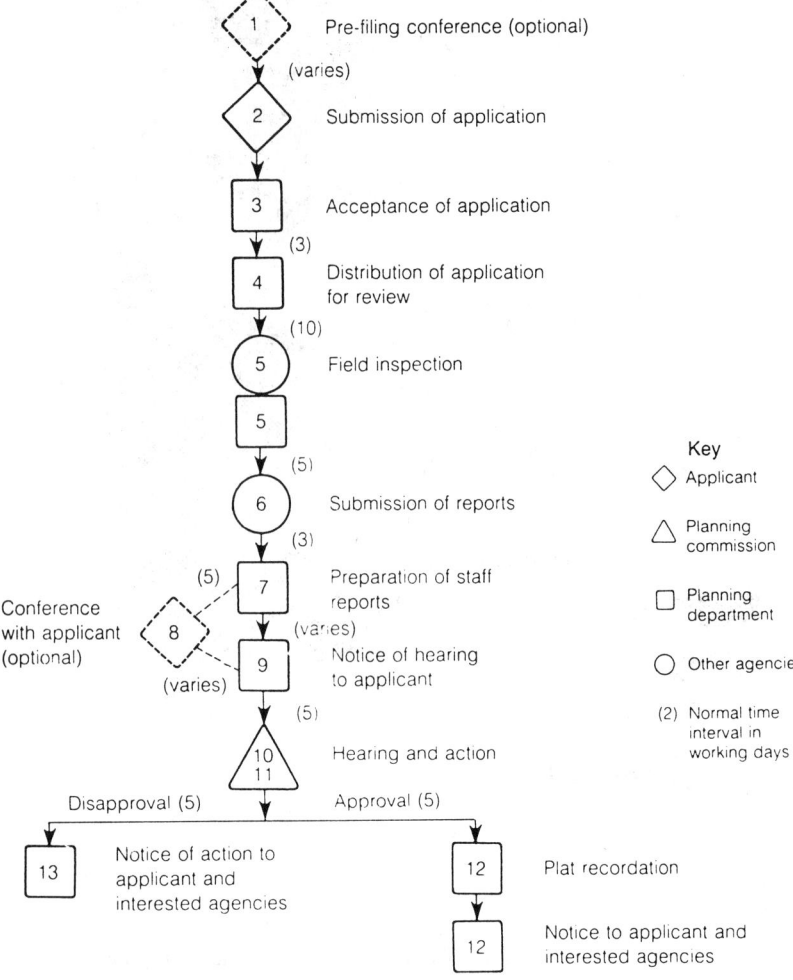

Figure 12. Our Town Final Plat Review Process. *Source*: Frank S. So and Judith Getzels, "Land Subdivision Regulation," in *The Practice of Local Government Planning (Second Edition)*, Frank S. So and Judith Getzels, eds. Washington, D.C.: International City Management Association, 1988, p. 232.

Example 12

Figure 12, shows that flow diagrams are useful for depicting step-by-step processes, especially when a decision may lead to any one of several alternative actions. The steps in the process can be keyed to a legend for additional information about activities at that step, optional steps can be indicated by broken lines, and steps in the process can be annotated. ∎

Cartoons

Used sparingly, cartoons can be effective means for conveying data. Cartoons need not always be humorous; they can be used to present a complex idea more simply.

Example 13

When you have bad news to report, a humorous cartoon can sometimes help to relieve tension. Smiling faces can be used, for example, in a public presentation where the presenter circles the face that depicts the audience's response to a particular question. In Figure 13, the collective response of Our Town citizens is presented as a frowning face. The same basic rules apply for cartoons. Keep it simple and identify your sources. ∎

Figure 13. The Collective Opinion of Our Town Citizens about the Tax Levy Increase. *Source*: Our Town Sample Survey, 1991.

When preparing graphics, we are attempting to present the essence of a set of data in the form of a picture. Therefore, we need to present the data simply and clearly. Select the most important relationships to illustrate; present only one or two ideas per graphic. Remove superfluous information from the graphic.

Further Readings

Dandekar, Hemalata C. *The Planner's Use of Information: Techniques for Collection, Organization, and Communication.* Stroudsburg, Pennsylvania: Hutchinson Ross, 1982.

Patton, Carl V., and David S. Sawicki, *Basic Methods of Policy Analysis and Planning.* Englewood Cliffs, New Jersey: Prentice-Hall, Inc., 1986.

Schmid, Calvin F., *Statistical Graphics: Design Principles and Practices.* New York: John Wiley and Sons, 1983.

Schmid, Calvin F., and Stanton E. Schmid, *Handbook of Graphic Presentation, Second Edition.* New York: John Wiley and Sons, 1979.

Spear, Mary Eleanor, *Charting Statistics.* New York: McGraw-Hill Book Co., 1952.

du Toit, S. H. C., A. G. W. Steyn, and R. H. Stumpf, *Graphical Exploratory Data Analysis.* New York: Springer-Verlag, 1986.

Tufte, Edward R., *The Visual Display of Quantitative Information.* Cheshire, Connecticut: Graphics Press, 1983.

Tufte, Edward R., *Envisioning Information.* Cheshire, Connecticut: Graphics Press, 1990.

Witzling, Lawrence, and Robert Greenstreet, *Presenting Statistics: A Manager's Guide to the Persuasive Use of Statistics.* New York: John Wiley and Sons, 1989.

PART 2
Basic Ways to Analyze Data

Chapter 4
SCATTERPLOTS ....

Definition

Scatterplots are simple graphic techniques that are often valuable for discovering trends, patterns, and relationships in data. Scatterplots are also called scattergrams or scatter diagrams. Scatterplots are graphic representations of two variables: one is measured on the y (vertical) axis and the second variable is measured on the x (horizontal) axis. Scatterplots assume that both variables are measured on an interval or ratio scale (see Chapter 1, Descriptive Statistics). Scatterplots may be modified to include additional information.

Scatterplots are strongly recommended before doing any tests for associations or relationships between variables. A scatterplot can help us determine whether two variables are related. For instance, a city could use a scatterplot to find out if the level of noise in downtown parks and plazas is related to the use of those parks and plazas.

Method

Use graphic techniques in order to prepare a scatterplot as a first step in determining if noise levels and use of the parks are related. The procedure is as follows:

43

Step 1. Establish and label the *x* and *y* axes. The *x* axis, the abscissa or horizontal axis, is assigned to average noise levels. By convention, the variable that is thought to be the "independent variable," is assumed to be the one that causes or explains the variation in the "dependent variable." The *y* axis, the ordinate or vertical axis, is by convention assigned to the dependent variable. In this example, we assign use of the parks to the *y* axis on the assumption that use may be in some sense dependent on the levels of noise (note that the opposite assumption is also reasonable in this example).

Step 2. Scale the *x* and *y* axes. The scatterplot requires sufficient detail in the scale of the axis for you to plot the data accurately (see Figure 1). We decide that average noise level can have a scale with labeled increments of 10 dB(A) so that the *x* (horizontal) axis does not become too cluttered. Person hours of use can be plotted on the *y* (vertical) axis with a scale increment of 100 hours.

Step 3. Plot the data. Each of the ten parks included in Table 1 has a single point on the scatterplot that is the intersection of its values for hours of use and average noise level. For instance, park number 1 is the point on the scatterplot at the intersection of noise level "81" and hours of use "200."

Example 1

Data on ten parks are presented in Table 1. ∎

Table 1

Noise Level and Usage in Urban Parks

Park or Plaza ID Number	Average noise Level in dB(A)	Person Hours of Use
10	58	600
9	80	100
8	48	400
7	29	900
6	75	300
5	60	400
4	62	500
3	38	700
2	43	800
1	81	200

Source: Data developed for examples.

Scatterplots

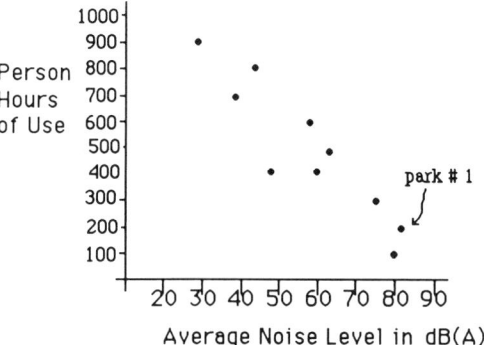

Figure 1. Scatterplot of Usage and Noise in Downtown Parks. *Source*: Data developed for example.

The scatterplot in Figure 1 provides useful information about noise levels and park usage. It clearly reveals an inverse (negative) relationship (see Figure 4). As noise levels increase, usage of the parks decreases. A scatterplot cannot tell us causality, i.e., whether noise causes the level of park use or whether the level of park use causes the level of noise measured. Causality should be established by use of theory, logic, research, or observation. The scatterplot in Figure 1 suggests that as noise levels increase there is a dramatic decrease in the use of the park. We would have to use statistical procedures to determine if this apparent relationship is statistically significant. A test of statistical significance (see Chapter 6, Statistical Significance, for further discussion) will tell us if we can be confident of the relationship or if the apparent relationship could easily be the result of sampling error. With only 10 parks in our sample data, we can not have a lot of confidence in our results. In our sample, the inclusion of one noisy park with high use could easily change the results of a statistical test for relationships.

The scatterplot also suggests the form of the relationship. The form of the relationship is important in deciding which statistical test to use. Figure 2 shows a direct (positive) linear relationship. This scatterplot presents a perfect linear relationship in which a straight line, with the x variable as the independent variable, perfectly predicts the value of the y variable. This means that a single straight line can connect all of the data points. In a direct relationship, as values of the x variable increase,

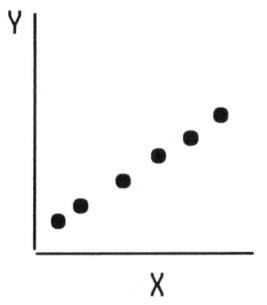

Figure 2. Perfect Direct (Positive) Linear Relationship.

Figure 3. Direct (Positive) Linear Relationship.

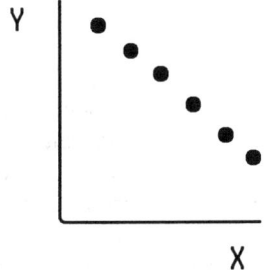
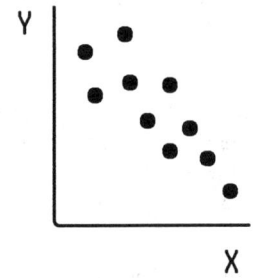

Figure 4. Perfect Inverse (Negative) Linear Relationship.

Figure 5. Inverse (Negative) Relationship.

the values of the y variable also increase. In Figure 3, a direct (positive) linear relationship is presented, but there is variation around any best fitting line through the plot of the data. See Chapter 5, Correlation Analysis for a discussion of quick methods of computing correlation.

Figure 4 presents a perfect inverse (negative) linear relationship. In an inverse relationship, as values of the x variable increase, values of the y variable decrease. Figure 5 presents an inverse (negative) relationship in which there is variation.

Scatterplots can also tell us if there is no relationship, or if the relationship that exists is curvilinear. Figure 6 presents a scatterplot of two variables that are not related. Knowledge of the x variable does not help us predict the value of the y variable. Figure 7 presents a curvilinear relationship, which in this example is an exponential relationship. A statistical test that assumes a linear relationship is unlikely

Scatterplots

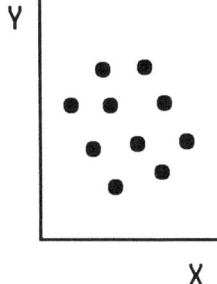

Figure 6. No Relationship (No Association).

Figure 7. Curvilinear (Exponential) Relationship.

to find or to determine accurately the strength of a relationship if the form of the relationship is curvilinear.

The scatterplots in Figures 1 through 5 reveal that the relationship is reasonably linear. Linear regression analysis would therefore be the preferred statistical technique. Linear regression analysis could be used to test for a relationship between use and noise levels in downtown parks in this city. The results of such an analysis reveal a best fitting line of $\hat{y} = 1255.36 - 13.33x$, $R^2 = 0.84$, and $S^2 = 12022.78$ (note: $\hat{y} = 1255.36 - 13.33x$ is the formula of the best fitting line for the data, and both R^2 and S^2 are measures of how well the regression line fits the data and the strength of the relationship.) Quick analysis techniques may also be used to measure the strength of the relationship between noise levels and park use (see Chapter 5, Correlation Analysis).

The scatterplot also reveals to us information about the distribution of the data. In using statistical tests to determine if there is a relationship between variables, it is important to know about the distribution of the data. Is there a single distribution, or are outliers present? An *outlier* is an observation that comes from a different population than the rest of the observations in the data set (this implies that the outlier is really from a different distribution). If one large rural park is included in the data on downtown parks, it is a statistical outlier that could drastically influence the results of a statistical test for a relationship (see Figure 8). The scatterplot helps identify outliers. Any data point on a scatterplot that is far removed from the rest of the data points should be investigated (Barnett and Lewis, 1978).

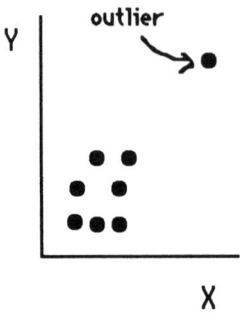

Figure 8. Scatterplot Showing an Outlier.

Further Readings

Barnett, Vic, and Toby Lewis, *Outliers in Statistical Data*. New York: John Wiley and Sons, 1978.

Blalock, Hubert M., Jr., *Social Statistics, Revised Second Edition*. New York: McGraw-Hill Book Company, 1979.

Draper, Norman R., and Harry Smith, *Applied Regression Analysis*. New York: John Wiley and Sons, 1966.

Krueckeberg, Donald A., and Arthur L. Silvers, *Urban Planning Analysis: Methods and Models*. New York: John Wiley and Sons, 1974.

Tufte, Edward R., *The Visual Display of Quantitative Information*. Cheshire, Connecticut: Graphics Press, 1983.

Chapter 5
CORRELATION
ANALYSIS . . . .

Definition

Correlation coefficients are used to determine whether a relationship exists between sets of two variables. In other words, measures of correlation help us determine whether there is a pattern to a set of data. When examining two variables, one independent and one dependent (i.e., the predictor and the predicted variables), a useful question to ask is, "How strong is the relationship between these variables?" Correlation coefficients give a numerical value that summarizes the strength and direction of the relationship between two variables.

Various correlation coefficients are used to analyze data, depending on whether the data are nominal, ordinal, interval, or ratio in scale. In quick analysis, *gamma* or *Yule's Q* is a useful correlation measure when data can be structured in ordinal form (Davis, 1971). This is the case with most public policy data, especially survey data. Gamma can also be applied to higher level (interval or ratio) data that have been categorized as ordinal data, e.g., income data that have been grouped into high, middle, and low income categories. Gamma is especially useful when data are available in tabular form and we wish to determine quickly whether there is a correlation between the variables.

Method

Gamma is a correlation coefficient that ranges from −1.0 to +1.0, with 0.0 indicating no relationship between the variables, +1.0 indicating a perfect positive relationship, and −1.0 indicating a perfect negative relationship. Since the data are ordered, that is, for each variable its values are scaled from low to high, a positive correlation means that the values of the variables are consistent: high values on one variable are associated or correlated with high values on the other variable. A negative correlation means that high values on one variable are associated with low values on the other variable.

Most of us already have a general understanding of correlation or association. For example, we probably believe that people with more years of education earn higher incomes. This, then, would be described as an hypothesis that asserts that there is a positive correlation between education and income. A correlation coefficient can be used to tell us how strong this association happens to be, if it indeed exists. In Chapter 2, Tabular Analysis, we learned about the correlation between age and preference for night baseball. A correlation coefficient can tell us the strength and direction of this relationship. Gamma is one of several measures of correlation that can be used for this purpose. We propose its use because it can be computed quickly with paper and pencil and its meaning is easily understood and conveyed.

Gamma is most easily computed from tabular data. Assume we have two variables, each with two values. The data should be laid out as in Table 1, where the letters a, b, c, and d simply label the cells. If we were doing this analysis for the night baseball example that is discussed in Chapter 2, Tabular Analysis, variable one might be *Age* and variable two would be *Approval or Disapproval of Night Baseball.*

Table 1

The Layout for Gamma

Variable One	Variable Two	
	Low	High
High	c	b
Low	a	d

Correlation Analysis

The value of gamma is determined by computing the relationship between the number of pairs of observations having the same ranking on the two variables and the number of pairs of observations having the opposite rankings on the two variables. The relationship is as follows:

$$\text{Gamma} = \frac{(\text{same} - \text{opposite})}{(\text{same} + \text{opposite})}.$$

The observations falling into the cells identified above as a and b have the same ranking (high on both variables and low on both, respectively). The observations falling into cells c and d have the opposite ranking (a high value on one variable and a low value on the other). The following formula can be used to compute gamma, where a, b, c, and d are replaced by the number of observations in the respective cells. (The symbol for Gamma is the Greek letter γ.)

$$\text{Gamma} = \frac{(a \times b) - (c \times d)}{(a \times b) + (c \times d)}.$$

Example 1

Assume data are being presented to the city council about citizens who support and oppose installing a new concert stage in the city park. We are told that 51% of the people living near the park support installing the stage. We discover, however, that the data were collected for both homeowners and renters. Consequently, the data can be structured as shown in Table 2.

Table 2

Resident Attitudes about the Concert Stage

Respondents' Homeownership Status	Position on Stage	
	Against	For
Renter	46	160
Owner	150	44
Total	196	204

Source: Sample survey of residents.

When we examine the data in this way, we immediately see that homeownership is related to one's position toward installing the stage. Gamma tells how strong the relationship is:

$$\text{Gamma} = \frac{(150 \times 160) - (46 \times 44)}{(150 \times 160) + (46 \times 44)}$$

$$= \frac{24{,}000 - 2{,}024}{24{,}000 + 2{,}024} = \frac{21{,}976}{26{,}024} = +.84.$$

The gamma of +.84 indicates a very strong positive relationship between home ownership status and one's position toward installing the concert stage. Although the majority of respondents favored installing the stage, homeowners, while in the minority in the area, overwhelmingly oppose the installation of the stage. Such data and their interpretation would likely be of great interest to city council members.

How do you decide whether a correlation is low, moderate, or high? The terms used to describe various numerical values are arbitrary and depend upon the conventions developed in various fields. James Davis (1971) has proposed the categories given in Table 3.

Table 3

Terms for Values of Gamma

Value of Gamma	Appropriate Phrase
+.70 or higher	A very strong positive association
+.50 to +.69	A substantial positive association
+.30 to +.49	A moderate positive association
+.10 to +.29	A low positive association
+.01 to +.09	A negligible positive association
.00	No association
−.01 to −.09	A negligible negative association
−.10 to −.29	A low negative association
−.30 to −.49	A moderate negative association
−.50 to −.69	A substantial negative association
−.70 or lower	A very strong negative association

Source: James A Davis, *Elementary Survey Analysis*, Prentice-Hall, Inc., Englewood Cliffs, New Jersey, 1971, p. 49.

Correlation Analysis

There are, of course, other conventions that could be used to name the levels of correlation, and you should use the convention that exists in your field. The most important point is to adopt a convention and stick to it so that you describe the same correlation level consistently in your work. ■

Example 2

In this example we investigate the relationship between income status and the percentage of family income spent on health care in a state. We construct Table 4 from reported data.

In this instance, gamma indicates that there is a low negative correlation between income and the percentage of income spent on health care. That is, households with lower incomes tend to spend a greater proportion of their incomes on health care.

Because the gamma statistic varies between -1.0 and $+1.0$, we can compare the strength of correlation among various tables. For the second example above, we could compare the association between income and health care expenditures among several states by examining the signs and strengths of the gammas for the relevant data for the various states. The correlation data would then be summarized as in Table 5. ■

Table 4

Household Expenditures on Health Care

Household Annual Income Level	Number of Households Spending Various Percentages of Income on Health Care	
	Less than 7.5%	7.5% and Above
$18,000 and Greater	110,000	100,000
Less than $18,000	40,000	50,000

Source: State Department of Health Annual Report.

$$\text{Gamma} = \frac{(40{,}000 \times 100{,}000) - (110{,}000 \times 50{,}000)}{(40{,}000 \times 100{,}000) + (110{,}000 \times 50{,}000)} = -.16.$$

Table 5

The Poor Tend to Pay More for Health Care

State	Correlation between Household Income and Health Care Expenditures
D	−.35
C	−.20
Our State	−.12
B	+.08
A	+.14

Sources: Health expenditure reports for selected states.

Example 3

In Chapter 4, data were given for a scatterplot of noise level and usage in urban parks. Those data can be collapsed into a 2 by 2 table, and gamma can be computed to determine a correlation coefficient. The original data are reported again in Table 6.

These data can be collapsed into a table, using the mid-point of both ranges as the dividing line. The results are shown in Table 7.

Table 6

Noise Level and Usage in Urban Parks

Park or Plaza ID Number	Average Noise Level Measured in dB(A)	Person Hours of Use
10	58	600
9	80	100
8	48	400
7	29	900
6	75	300
5	60	400
4	62	500
3	38	700
2	43	800
1	81	200

Source: Data developed for the example in Chapter 4, Scatterplots.

Table 7
Noise Level Affects Park Usage

Person Hours of Use	Average Noise Level in dB(A)	
	0–49	50+
500+	3	2
0–499	1	4

$$\text{Gamma} = \frac{(1 \times 2) - (3 \times 4)}{(1 \times 2) + (3 \times 4)} = -.71.$$

This indicates that there is a very strong negative correlation between person-hours of park use and average noise level. That is, the parks with greater noise levels tend to have less use. ∎

Cautions and Extensions

While gamma has a number of properties that make it very useful for quick analysis, it does have some limitations that must be recognized. If there are no observations in one cell of a simple two by two table, the computed value will be a perfect positive or negative correlation. Thus, when this condition occurs, gamma should not be used as the correlation measure. When there is a small number of observations in one or more cells of any sized table the computed value should be interpreted cautiously as it may be overstating the relationship. A good rule of thumb would be not to use gamma when one or more of the cells contains fewer than five observations.

Gamma can be computed for variables with more than two values, and gamma can be extended to partial correlation analysis (Davis, 1971). After you have had some experience with tabular analysis and gamma, you will want to explore these possibilities. Moreover, the question of statistical significance for gamma has not been addressed here (Freeman, 1965). These issues require more than back-of-the-envelope calculations.

Finally, we are suggesting that you use gamma to analyze nominal data which are categorized into two categories that are then considered as ordinal data, and to analyze interval data that have been grouped as ordinal data. If sufficient time were available for a more complicated analysis, the experienced analyst would not take continuous data, like income and spending, as shown in Table 4, and consider it as grouped data, because too much information is lost in this process. There are statistics that may be more appropriate for use in these cases which the more experienced analyst might wish to apply. For example, Pearson's r, or the Pearson product moment correlation coefficient, would be a better measure of correlation for interval or ratio data, although it is more complicated and difficult to compute. *Lambda* or *chi square* (see Chapter 6, Statistical Significance) might also be considered as measures for nominal data.

All in all, we believe gamma is the most useful measure of correlation for *quick* analysis. You may wish, however, to consult a basic statistics book about the other statistical measures.

Further Readings

Blalock, Hubert M., Jr., *Social Statistics, Revised Second Edition.* New York: McGraw-Hill Book Co., 1979.

Davis, James A., *Elementary Survey Analysis.* Englewood Cliffs, New Jersey: Prentice-Hall, Inc., 1971.

Freeman, Linton D., *Elementary Applied Statistics: For Students in Behavioral Science.* New York: John Wiley and Sons, 1965.

Patton, Carl V., and David S. Sawicki, *Basic Methods of Policy Analysis and Planning.* Englewood Cliffs, New Jersey: Prentice-Hall, Inc., 1986.

Runyon, Richard P., and Audrey Haber, *Fundamentals of Behavioral Statistics, Fourth Edition.* Reading, Massachusetts: Addison-Wesley, 1980.

Wonnacott, Ronald J., and Thomas H. Wonnacott, *Statistics: Discovering Its Power*, New York: John Wiley and Sons, 1982.

Wonnacott, Thomas H., and Ronald J. Wonnacott, *Introductory Statistics for Business and Economics, Third Edition.* New York: John Wiley and Sons, 1984.

Chapter 6
STATISTICAL SIGNIFICANCE

Definition

Sometimes people refer to *significant findings*. They often mean that what they discovered was interesting. When statisticians refer to a significant finding they mean that the interesting thing discovered most likely *did not occur by chance*, or, for example, that only 5 times out of 100 would the relationship occur by chance. Measures of association, such as gamma, discussed in Chapter 5, tell us whether two or more variables are correlated and the strength of that correlation. *Measures of significance* tell us the probability that the association occurred by chance. *Chi square* (χ^2) is a useful measure of significance. It can be used to examine data displayed in tabular form, and it is handy for quickly reexamining data presented by others.

If we were to flip a coin 100 times we would expect to get approximately 50 heads and 50 tails. If we got a 60/40 split we would not be disturbed because in the short run we expect some variation. If we flipped the coin 1,000 times we would expect results fairly close to a 500/500 split. If we found an 800/200 split we would expect that something other than chance was involved. We would likely expect a biased coin. If we were flipping the coin as part of a betting event, we would certainly want to look at the coin. We would want to look at the coin because what we observed, the 80%/20% split, was not what we expected, a 50%/50% split. We would think that the outcome of the coin flipping cannot be explained on the basis of chance alone. The concept of *comparing observed values with expected values* is the basis for the χ^2 statistic.

Method

To compute χ^2 we compare the distribution of observed values for a set of data with the values we expected, had there been no association between the variables. Assume that we wanted to examine the drinking preferences of professional boxers. If being a crowned champion or merely a contender had no bearing on one's milk preference, we would expect the same proportion of a sample of boxers to prefer homogenized or skimmed milk, as shown in Table 1, assuming in general a 50%/50% preference for skimmed and homogenized milk.

The expected cell frequencies are obvious in this simple example, but for more complicated tables, they are computed using the respective row and column totals for each cell (the row and column marginals).

$$\text{ECF (expected cell frequency)} = \frac{\text{row marginal} \times \text{column marginal}}{\text{total number of observations}}.$$

Table 1

Expected Milk Preferences

Fighter Status	Milk Preference		Total in Sample
	Skimmed	Homogenized	
Champion	20	20	40
Contender	30	30	60
Totals	50	50	100

Source: Analyst estimate.

Note: To produce this table all we had to know was the number of champions and contenders and the ratio between people who prefer skimmed and homogenized milk. We then estimated the four cell values. The data in this table are raw frequencies, not percentages.

Statistical Significance

The expected frequency for the contender/skimmed cell would be computed:

$$\text{ECF} = \frac{60 \times 50}{100} = 30.$$

If we took a sample of 100 milk drinking boxers and found that the preferences tended to group into two quadrants rather than being distributed among all four cells, we would expect that there is some relationship (correlation or association) between fighter status and milk preference. (We could compute a measure of correlation or association, but because we have taken a sample, we would also need to compute a measure of significance to determine whether the association occurred simply because of chance.) If there is a close correspondence between the expected and the observed values, this would be an indication that there is no relationship between fighter status and milk preference.

Example 1

Assume that we conducted a random sample of U.S. boxers, in which 40% of the boxers are at the championship level and 60% are contenders, and found the milk preferences given in Table 2.

This would indicate that fighter status and milk preference are correlated, that champions prefer homogenized milk. In fact, we obtain a gamma of .71, but are the findings significant? Remember, we are asking whether these results could have occurred by chance, not

Table 2

Observed Milk Preferences

Fighter Status	Milk Preference		Total in Sample
	Skimmed	Homogenized	
Champion	10	30	40
Contender	40	20	60
Totals	50	50	100

Source: Sample survey.

whether they are interesting, nifty, or fantastic. The χ^2 formula compares expected and observed cell frequencies to measure the difference between observed and expected values:

$$\chi^2 = \sum \frac{(O-E)^2}{E},$$

where

O = observed cell frequencies,
E = expected cell frequencies, and
Σ = summation of the results of the following computation.

That is, χ^2 equals the observed first cell frequency minus the expected first cell frequency squared, divided by the expected first cell frequency, plus the same for the second cell, etc. Chi square must be calculated from absolute numbers, not percentages.

For the foregoing data:

$$\chi^2 = \frac{(10-20)^2}{20} + \frac{(30-20)^2}{20} + \frac{(40-30)^2}{30} + \frac{(20-30)^2}{30}$$

$$= \frac{(-10)^2}{20} + \frac{(10)^2}{20} + \frac{(10)^2}{30} + \frac{(-10)^2}{30}$$

$$= \frac{100}{20} + \frac{100}{20} + \frac{100}{30} + \frac{100}{30}$$

$$= 5.0 + 5.0 + 3.33 + 3.33$$

$$= 16.66.$$

The χ^2 value of 16.66 is interpreted by finding its location in a list of values that tells us, for a table with a certain number of cells, the probability that a χ^2 value as large or larger than the one computed would occur by chance. The value is also affected by the *degrees of freedom*, which is the number of quantities that are unknown minus the number of independent equations linking these unknowns. For data in a contingency table, this boils down to finding out how many cells you must have filled in before you can determine all the cells. For a two by two table, where you know the row and column marginals, you need to know the value in only one cell to figure out all the other cells. Prove this to yourself for the milk preference example. For a three by three

Statistical Significance

table you can fill in the last row and column if you know two rows and two columns. The formula for degrees of freedom (d.f.) for a contingency table is: d.f. = $(r - 1)(c - 1)$, or the product of the number of rows (r) minus one and the number of columns (c) minus one.

Certain levels of significance have been adopted for use in statistical analysis. In medical research, a very small chance of the sample not representing the population is essential. Thus, findings are considered statistically significant only if they would have occurred by chance no more than once out of 100 samples. Such results are said to be statistically significant at the .01 level. In social science research, the traditional significance level is .05, that is, statistically significant results are those that would occur by chance in no more than 5 out of 100 samples. Theoretically, statistical significance applies to say, 5 out of 100 samples. In practice, we refer to the chance that values in a particular sample (usually we take only one) are the same as those in the population. Significance at the .05 level means that 5 out of 100 times the differences observed may have occurred by chance; it does not mean that the differences observed are accurate within 5% of the population value.

Knowing that there is one degree of freedom for our milk preference problem, we can look up the χ^2 value of 16.66 in a χ^2 table (see Table 8 at the end of this chapter). We look up the χ^2 value by first selecting a significance level (e.g., .05), then finding the value for the degrees of freedom (i.e., 1). We then check to see if the computed value is greater than the table value. Not only is our computed value larger than the table value for one degree of freedom at the .05 level (3.84), but our value of 16.66 is so large that it cannot even be found for one degree of freedom in a regular table. This indicates that a difference between observed and expected as large as we found would occur less than once in one thousand samples. Consequently, we say the relationship is statistically significant. It also may be interesting or important. ∎

Example 2

Consider the data in Table 3, which a venture capitalist has prepared based on a sample of attitudes of old and new city residents toward industry.

Table 3

Observed Data on Attracting Industry

Residents	Con Industry	Pro Industry	Total
Old	5	20	25
New	15	10	25
Totals	20	30	50

Source: Venture capitalist sample.

We want to know whether old and new residents hold different opinions. In statistical language, the *null hypothesis* is that there is no relationship between the two variables. In other words, we are saying that there will be negligible or no difference between the observed and expected values. We first compute a table of expected cell frequencies based on the marginal frequencies, using the ECF formula. For the old residents/con industry cell, we would compute the expected cell frequency as:

$$\text{ECF} = \frac{25 \times 20}{50} = \frac{500}{50} = 10.$$

The expected values are shown in Table 4.

Table 4

Expected Data on Attracting Industry

Residents	Con Industry	Pro Industry	Total
Old	10	15	25
New	10	15	25
Totals	20	30	50

Source: Analyst estimate.

Statistical Significance

Chi² or χ^2 is used to summarize the extent of difference of the observed from the expected figures,

$$\chi^2 = \sum \frac{(O-E)^2}{E}$$

$$= \frac{(5-10)^2}{10} + \frac{(20-15)^2}{15} + \frac{(15-10)^2}{10} + \frac{(10-15)^2}{15}$$

$$= \frac{(-5)^2}{10} + \frac{(5)^2}{15} + \frac{(5)^2}{10} + \frac{(-5)^2}{15}$$

$$= 2.5 + 1.667 + 2.5 + 1.667 = 8.33.$$

degrees of freedom = (number of rows − 1) (number of columns − 1);

$$\text{d.f.} = (2-1)(2-1) = 1.$$

For significance at the .05 level, we need a χ^2 value of at least 3.84 (see Table 8). Thus, the null hypothesis is rejected. (The value of 8.33 we computed for the given data exceeds the table value of 3.84, so we reject the null hypothesis of no relationship.) Given the marginal distribution of the variables, an observed table at least as different, or more different, than the expected table would occur by chance less than one time in one hundred. This is depicted graphically in Figure 1, where the probability of obtaining a χ^2 value of 8.33 is shown to be very small, that is, such a large value occurs infrequently by chance: ■

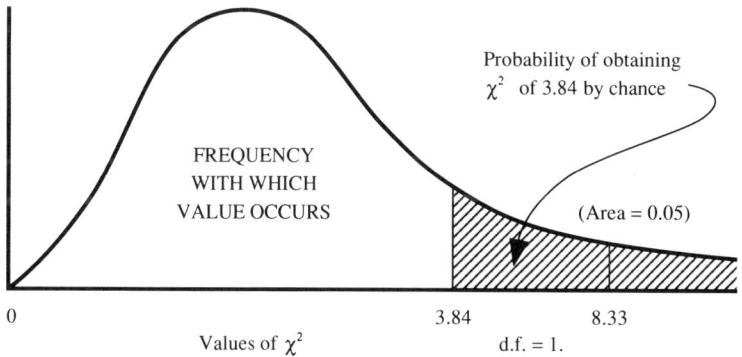

Figure 1. The Probability of Obtaining a χ^2 of 3.84 by Chance.

Example 3

Chi square can also be computed for tables larger than two by two, as shown in Tables 5a and 5b. For example:

Table 5a			
Observed Values			
113	60	27	200
31	91	38	160
8	8	24	40
152	159	89	400

Table 5b			
Expected Values			
76	81	44	200
61	63	36	160
15	16	9	40
152	159	89	400

Source: Analyst data files.

In this case it is not easily apparent whether there is a significant difference between the observed values and the expected. So, we ask the question whether there is a statistically significant difference at the .05 level.

$$\chi^2 = \sum \frac{(O-E)^2}{E}$$

$$= \frac{(113-76)^2}{76} + \frac{(60-81)^2}{81} + \frac{(27-44)^2}{44}$$

$$+ \frac{(31-61)^2}{61} + \frac{(91-63)^2}{63} + \frac{(38-36)^2}{36}$$

$$+ \frac{(8-15)^2}{15} + \frac{(8-16)^2}{16} + \frac{(24-9)^2}{9}$$

$$= \frac{1369}{76} + \frac{400}{80} \text{ etc.}$$

$$= 18 + 5.44 + 6.6 + 14.75 + 12.4 + .11 + 3.3 + 4 + 25 = 89.6.$$

d.f. $= (r-1)(c-1) = 4$.

Statistical Significance

At .05 a value of at least 9.49 is required for four degrees of freedom (Table 8). Thus, we can infer that the difference did not likely occur by chance. Thus, the null hypothesis of no relationship is rejected. ■

Significance and Sample Size

Be aware that increasing the sample size will affect the significance level. (See also Chapter 13, Sample Size.) Significance is used to describe the chance that a sample does not represent the population from which it was taken. A small sample, then, has less chance of representing a population than a large sample. A sample of 70 households from a town of 10,000 households is not likely to be as representative as a sample of 700 households.

The larger the sample the greater the chance that the values it describes are the same as those in the population, assuming that both samples were selected with equal care and precision. This should become evident when we think about taking larger and larger samples until we reach the point when our sample is no longer a sample but a census of the entire population. Thus, the larger the sample the more likely it will be significant, that is, the more likely the values from the sample will be the same as those in the population. Although findings might be statistically significant, they may not be important.

Example 4

The example given in Tables 6a and 6b and Tables 7a and 7b shows that merely taking a larger sample will cause our findings to be statistically significant at a higher level, that is, have less likely occurred by chance. The statistics show that the degree of association is the same but that the larger sample (Tables 7a and 7b) has a higher level of statistical significance, although the importance of the issue has not changed simply because a larger sample was taken.

66   Basic Ways to Analyze Data

Table 6a			Table 6b		
Observed Values			Expected Values		
5	20	25	7	18	25
15	30	45	13	32	45
20	50	70	20	50	70
Gamma = .33			Gamma = .02		

Sample of 70.

$$\chi^2 = \frac{(5-7)^2}{7} + \frac{(20-18)^2}{18} + \frac{(15-13)^2}{13} + \frac{(30-32)^2}{32}$$
$$= 1.23$$
$$\text{d.f.} = 1.$$

Thus, with a sample of 70 the relationship is not significant at the .05 level.

Table 7a			Table 7b		
Observed Values			Expected Values		
50	200	250	70	180	250
150	300	450	130	320	450
200	500	700	200	500	700
Gamma = .33			Gamma = .02		

Sample of 700.

$$\chi^2 = \frac{(50-70)^2}{70} + \frac{(200-180)^2}{180} + \frac{(150-130)^2}{130} + \frac{(300-320)^2}{320}$$
$$= 12.25$$
$$\text{d.f.} = 1$$

Thus, with a sample of 700 the relationship is significant at the .05 level. In fact, the difference would occur by chance less than once out of 1,000 samples. ∎

Interpreting Chi Square

Chi square does not imply causation, but indicates the probability that the relationship occurred by chance. Furthermore, χ^2 does not tell us the meaning of the association. It must be interpreted from other facts. While χ^2 is easy to use and to interpret, caution must be exercised when the table cells have low expected frequencies (e.g., less than five expected observations per cell). It also is designed for nominal and ordinal data. For interval data the F distribution test might be more appropriate. Note also that when χ^2 is used for a 2 by 2 table with a small number of cases, you should use a correction factor or the Fisher's exact test (Blalock, 1979, Section 15.2). But even with these limitations, the χ^2 test is a useful measure of statistical significance that can be applied quickly and that will provide useful results.

Interpreting Statistical Differences

We use tests of statistical significance to draw conclusions from the data we are examining. In classical hypothesis testing we state a hunch or idea about a population that can be tested through a random sample. We usually state a *null hypothesis* (H_0) that says there is *no difference* between the sample value and the value in the population from which the sample was selected. We then apply a statistical test to the data in an attempt to disprove the null hypothesis. That is, if we calculate that the results of the statistical test are statistically significant, then we can reject the null hypothesis of no difference and assert the alternative hypothesis (H_1) that there is a difference. We cannot directly prove the alternative hypothesis. It is affirmed by *rejecting* or disproving the null hypothesis.

If the test statistic is not significant, this does not necessarily mean that we accept the null hypothesis of no relationship between the variables. The alternative to rejecting a null hypothesis *is not* simply the acceptance of the null hypothesis. The most we can claim is that we failed to reject the null hypothesis, which means we accept the null hypothesis as one of many plausible hypotheses.

In the real world the null hypothesis is either true or false. In making a decision to reject or not reject the null hypothesis, we always risk making a mistake. We can make an error by rejecting a true hypothesis (statisticians call this a *Type I* error), or we can error by accepting a false hypothesis (a *Type II* error) (Runyon, 1980, pp. 200–203). The only way one error can be reduced without increasing the other error is by collecting better evidence with which to test the hypothesis.

When we reject the null hypothesis, we are saying that the difference we found between the two variables cannot likely be attributed to chance or sampling error. But there is always the possibility that once in a while (e.g., 5 out of 100 times or 1 out of 100 times) a difference that seems to be present does not really exist. This leads us to remind the reader that caution must always be exercised when drawing conclusions from a statistical test alone. If a series of tests and other evidence all point to the same conclusion, the chance of making a mistake in interpretation is lessened.

Statistically significant differences alone, however, cannot prove cause and effect relationships. Rather, they suggest possibilities and help us identify directions for further study.

The p-Value Test as an Alternative Approach

In classical hypothesis testing we set a significance level or error level for inferring the impact of non-chance factors (see Chapter 12, Confidence Levels). That is, when the event would occur five percent of the time or less by chance, we are willing to assert that the results are most likely due to non-chance factors. This cut off point is known as the significance level, e.g., 5% significance or .05. This level is selected by researchers or analysts by using accepted conventions in their fields, for example at .05 for social science research and .01 for medical experiments. This level of significance set by the analyst is called the *alpha* (α) level. Recall that we commit a Type I error when we reject the null hypothesis when it is true. The probability of making the error is α. This means that when we set the rejection point at .05 significance we reject by mistake the null hypothesis 5% of the time. In this chapter we set the alpha value at 5%. This was shown graphically in Figure 1 as the area identified as 0.05 of the frequency distribution.

In making a Type II error we fail to reject the null hypothesis when it is false. *Beta* (β) is the probability of making a Type II error. This is the more common error. The lower we set the rejection level, the lower the chance of a Type I error and the greater the chance of a Type II error. We set the rejection level low to avoid claiming a result that is not true. It is usually better to fail to claim a true result than to claim a result that is not true. As an example, it is better to free a guilty person (Type II) than to condemn an innocent person (Type I).

Statistical Significance

Instead of setting the alpha value, we can compute what is called the *p-value*, which allows us to calculate the extent to which the null hypothesis is supported by the data. The *p*-value is defined as the probability that the sample value would be as large as the value of the statistic actually observed if the null hypothesis were true (Wonnacott and Wonnacott, 1984, p. 263). The *p*-value summarizes the degree to which the data support the null hypothesis. As such, it is a measure of the credibility of the null hypothesis. If the *p*-value is equal to or less than alpha, then the null hypothesis is rejected.

Today statisticians increasingly prefer *p*-values over classical testing because classical tests involve arbitrarily setting alpha, for example, as we did in this chapter, at 5%. Instead, the analyst states the computed *p*-value and lets the reader use this figure to decide on the validity of the null hypothesis. At this point we have not presented sufficient data to allow the reader to calculate a *p*-value, but we will return to the topic in Chapter 12, Confidence Levels.

Table 8

The Chi Square Distribution

χ^2 Critical Points

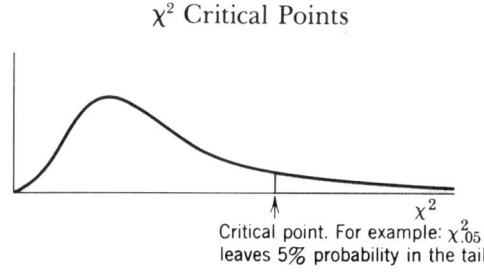

Critical point. For example: $\chi^2_{.05}$ leaves 5% probability in the tail.

d.f.	$\chi^2_{.25}$	$\chi^2_{.10}$	$\chi^2_{.05}$	$\chi^2_{.025}$	$\chi^2_{.010}$	$\chi^2_{.005}$	$\chi^2_{.001}$
1	1.32	2.71	3.84	5.02	6.63	7.88	10.8
2	2.77	4.61	5.99	7.38	9.21	10.6	13.8
3	4.11	6.25	7.81	9.35	11.3	12.8	16.3
4	5.39	7.78	9.49	11.1	13.3	14.9	18.5
5	6.63	9.24	11.1	12.8	15.1	16.7	20.5
6	7.84	10.6	12.6	14.4	16.8	18.5	22.5
7	9.04	12.0	14.1	16.0	18.5	20.3	24.3
8	10.2	13.4	15.5	17.5	20.1	22.0	26.1
9	11.4	14.7	16.9	19.0	21.7	23.6	27.9

(Cont)

Table 8
(Cont.)

d.f.	$\chi^2_{.25}$	$\chi^2_{.10}$	$\chi^2_{.05}$	$\chi^2_{.025}$	$\chi^2_{.010}$	$\chi^2_{.005}$	$\chi^2_{.001}$
10	12.5	16.0	18.3	20.5	23.2	25.2	29.6
11	13.7	17.3	19.7	21.9	24.7	26.8	31.3
12	14.8	18.5	21.0	23.3	26.2	28.3	32.9
13	16.0	19.8	22.4	24.7	27.7	29.8	34.5
14	17.1	21.1	23.7	26.1	29.1	31.3	36.1
15	18.2	22.3	25.0	27.5	30.6	32.8	37.7
16	19.4	23.5	26.3	28.8	32.0	34.3	39.3
17	20.5	24.8	27.6	30.2	33.4	35.7	40.8
18	21.6	26.0	28.9	31.5	34.8	37.2	42.3
19	22.7	27.2	30.1	32.9	36.2	38.6	32.8
20	23.8	28.4	31.4	34.2	37.6	40.0	45.3
21	24.9	29.6	32.7	35.5	38.9	41.4	46.8
22	26.0	30.8	33.9	36.8	40.3	42.8	48.3
23	27.1	32.0	35.2	38.1	41.6	44.2	49.7
24	28.2	33.2	36.4	39.4	32.0	45.6	51.2
25	29.3	34.4	37.7	40.6	44.3	46.9	52.6
26	30.4	35.6	38.9	41.9	45.6	48.3	54.1
27	31.5	36.7	40.1	43.2	47.0	49.6	55.5
28	32.6	37.9	41.3	44.5	48.3	51.0	56.9
29	33.7	39.1	42.6	45.7	49.6	52.3	58.3
30	34.8	40.3	43.8	47.0	50.9	53.7	59.7
40	45.6	51.8	55.8	59.3	63.7	66.8	73.4
50	56.3	63.2	67.5	71.4	76.2	79.5	86.7
60	67.0	74.4	79.1	83.3	88.4	92.0	99.6
70	77.6	85.5	90.5	95.0	100	104	112
80	88.1	96.6	102	107	112	116	125
90	98.6	108	113	118	124	128	137
100	109	118	124	130	136	140	149

Reprinted from Ronald J. Wonnacott and Thomas H. Wonnacott, *Statistics: Discovering Its Power*, New York: John Wiley and Sons, 1982, p. 352.

Further Readings

Blalock, Hubert M., Jr., *Social Statistics, Revised Second Edition*. New York: McGraw-Hill Book Co., 1979.

Caulcott, Evelyn, *Significance Tests*. London: Routledge and Kegan Paul, 1973.

Franzblau, Abraham, *A Primer of Statistics for Non-Statisticians*. New York: Harcourt, Brace and World, Inc., 1958.

Runyon, Richard P., and Audrey Haber, *Fundamentals of Behavioral Statistics, Fourth Edition*. Reading, Massachusetts: Addison-Wesley, 1980.

Wonnacott, Ronald J., and Thomas H. Wonnacott, *Statistics: Discovering Its Power*. New York: John Wiley and Sons, 1982.

Wonnacott, Thomas H., and Ronald J. Wonnacott, *Introductory Statistics for Business and Economics, Third Edition*. New York: John Wiley and Sons, 1984.

Part 3
Looking at Data Across Time

Chapter 7
PROJECTION TECHNIQUES ......

Definition

Projection techniques are quantitative methods for estimating the future. Techniques to project the future are essential for making plans and for most policy decisions.

There are no projection, prediction, or forecasting techniques that are guaranteed to be accurate. This chapter reviews two types of quick and easy to use techniques to project future levels of population, employment, economic output, or other activities based on past levels of those measurements or activities. These techniques are designed to provide reasonable projections of future levels of measurements or activities. If you are making mid-range or long-range projections (projections of more than 10 years), then there are no methods that are likely to be accurate and these quick methods may be fine. If you need to make short-term accurate projections (less than 10 percent error), then more accurate methods such as cohort-survival for population projections or input-output or econometric models for economic projections may be necessary.

Depending on your purpose, you may want to round off your projection to a level of accuracy commensurate to your degree of confidence in the projected number. For a projection two years into the future you might feel comfortable rounding to a value close to your calculated projection. For a long-term projection, you might want to round off to the nearest thousand or million, depending on the units involved and your confidence in your projection. You should be especially careful in making presentations of projection results not to present figures that imply an accuracy of projection with which you are not comfortable.

Method

The two types of simple projection techniques reviewed in this chapter are *trend extrapolation* projection models and *ratio* projection models. Deciding which of these techniques is best for your needs requires that you evaluate the data you have available and your assumptions. Good historical data are necessary for trend extrapolation projection models. These models assume that there will be no abrupt or dramatic changes from conditions affecting growth during the period of the historical data. Projections for a larger entity that includes the unit for which you are making projections are required for ratio projection techniques. These models assume that you are confident about projections available for the larger entity. One of the first steps in deciding which technique to use is to plot the data against time to observe any trends or patterns in the data (see time-series plots in Chapter 3, Graphic Techniques).

Trend Extrapolation Projection Models

Making projections into the future based on trends in the historical data is one commonly used projection technique. Using one of the trend extrapolation projection models requires that you first identify the trend in the historical data correctly. If the trend extrapolation model does not fit the historical data, then the resulting projections may be impossible to justify and are likely to be highly inaccurate. Standard statistical techniques are available for determining the variation when the model selected is fitted to the historical data. The amount of error present in fitting the model to the historical data provides a rough estimate of the potential for error in the projections.

A time-series plot of the historical data is an essential first step before making any projection. The basic shapes of trends in the histori-

Projection Techniques

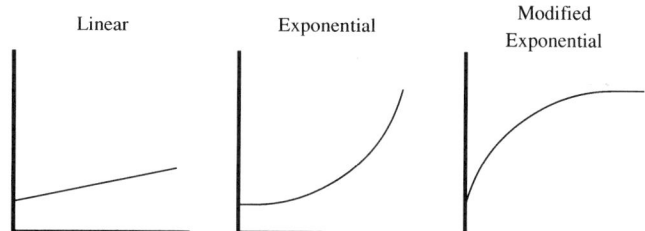

Figure 1. Generalized Shapes of Trends in Historical Data. *Source*: Developed for examples.

cal data are important in understanding the subject you are going to project, and in helping guide you in selecting the appropriate projection technique. Figure 1 illustrates three generalized forms of trends in historical data.

Actual data are unlikely to fit one of these generalized shapes exactly. If the data approximates one of these shapes you may find a trend extrapolation projection model to be reasonable and easy to use.

Sometimes all of the data are not included in the projection model. If the historical data extend far back in time, some of the oldest data may fit one model while the more recent historical data may fit a different model. Including only the more recent historical data in the projection model assumes that the trends of the recent past will continue. Many things that are projected have a cyclical pattern. Basing projections on only a portion of a cycle can produce dramatically inaccurate projections. When using trend extrapolation projection models to project cyclic phenomena, be sure to include sufficient historical data to establish a clear trend.

Before using a quantitative model, it is useful to graphically extend the trend. This form of a rough projection may be all that your purpose requires. If you use the following quantitative models, comparing their results to your graphic extension of the trend will be a useful check on the model results and their reasonableness.

Extrapolation projection models examine the historical data using a time-series plot. The historical data in Table 1 will be used in the examples of extrapolation projection techniques.

Table 1
Historical Data for Trend Extrapolation Examples

Year	(1) Linear	(2) Exponential	(3) Modified Exponential
1986	173,422	20,000	20,000
1987	172,246	21,000	29,000
1988	170,173	23,900	36,000
1989	170,056	28,000	37,500
1990	169,193	33,000	38,000

Source: Data developed for examples.

Linear Projection

A *linear (straight line) projection model* is appropriate for making projections when the subject has a history of nearly equal change for each time interval over the recent past. For example, if population has increased by 4000 people each year for the past ten years, it may be reasonable to use a linear model that assumes that population will continue to increase by 4000 people each year for the next few years. A variety of modifications to the basic model are available. A simple time-series plot of the data for population in previous years will quickly indicate if the past trend has been reasonably linear.

The mathematical formula for the linear model is

$$Y_{t+n} = Y_t + bn,$$

where

Y = the measurement or activity being projected,
t = the unit of time of the data,
Y_t = the most recent time interval of the historical data and the starting point for the projection,
n = the number of units of time (in months, years, etc.),
Y_{t+n} = the activity n units of time from t, and
b = the average amount of growth or decline per unit of time.

To use the linear model, we must calculate b, the average annual growth or decline amount. We must average the historical changes

Projection Techniques

from one time period to the next using the formula

$$b = \frac{\sum_{i=1}^{m}(Y_t - Y_{t-1})}{m,}$$

where

b = the average annual growth or decline increment per unit of time,
Σ = the addition of what follows; in this model we must add together all the annual changes from the first to the most recent ($i = 1$ to the mth) time period,
Y_t = the most recent time interval of the historical data,
Y_{t-1} = the level of Y one time period before Y_t, and
m = the number of historical intervals over which the average annual growth is calculated.

Example 1

A linear model is used to project the "linear" data provided in Table 1.

To determine if a linear model is appropriate for projecting this data, a time-series plot (see Figure 2) of the data titled "linear" in Table 1 is useful.

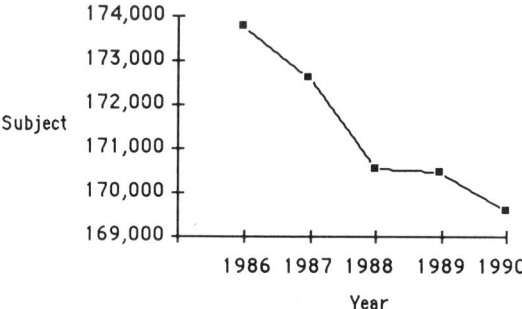

Figure 2. A Time-series Plot of the "Linear" Data from 1986 to 1990.
Source: Data developed for examples.

Table 2

Annual Changes in "Linear" Data

Year t	Year $t-1$	Difference
1987 minus	1986	= –1176
1988 minus	1987	= –2073
1989 minus	1988	= –117
1990 minus	1989	= –863

Source: Data developed for examples.
Sum of the changes from 1986 to 1990 = –4229.
Average change (b) equals –4229/4 = –1057.3.

There has been a relatively steady decrease from 1986 to 1990. If we make the assumption that this trend will continue, we can use a linear model to project this data into the future. Using our formula to project our subject one year into the future from 1990, our formula becomes:

$$\text{Subject}_{1991} = \text{Subject}_{1990} + b(1).$$

The term b in the model stands for the average growth or decline increment per unit of time. It is calculated, according to the formula presented above, by adding together the annual changes in our subject and dividing by the number of yearly changes (see Table 2).

Two projections for the "linear" data from 1986 using the linear model are:

Subject in 1991 = 169,193 + (–1057.3)(1) = 168,135.8 = 168,136.
Subject in 1993 = 169,193 + (–1057.3)(3) = 166,021.3 = 166,021.

■

Exponential Projection Model

The *exponential projection model* may be appropriate when the rate of change remains constant over the intervals of time in the historical data. An example might be a town where the population increases eight percent each year. If the percentage change in population growth remains constant over time, then the absolute number of new people will be constantly increasing because eight percent of a larger population will be a larger number of people in each of the following years.

Projection Techniques

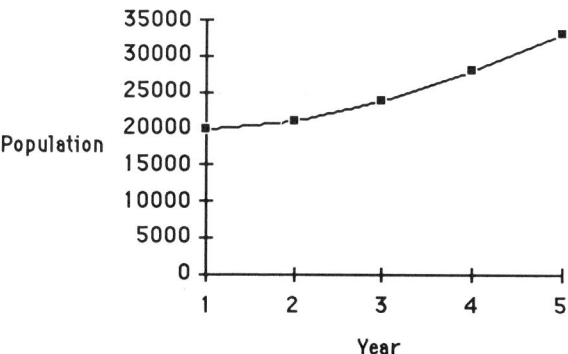

Figure 3. Time-series plot of exponential data. *Source*: Data developed for examples.

When a time-series plot and examination of the historical data reveal that the subject to be projected has experienced a pattern of a constant rate of change for the increments of time, then an exponential projection model may be an appropriate technique. A time-series plot of the data labelled "exponential" in Table 1 is presented in Figure 3.

The mathematical formula for the exponential model is:

$$Y_{t+n} = Y_t(1+r)^n$$

$$r = \frac{1}{m} \sum \frac{Y_t - Y_{t-1}}{Y_{t-1}}$$

where

    Y = the measurement or activity being projected,
    t = the unit of time of the data,
    Y_t = the most recent time interval of the historical data and the starting point for the projection,
    n = the number of units of time (in months, years, etc.),
    Y_{t+n} = the activity n units of time from t,
    r = the rate of growth or decline from the beginning to the end of the historical data,
    Y_{t-1} = the level of Y one time period before Y_t, and
    m = the number of historical intervals over which the average annual growth is calculated.

Table 3
Annual Rate of Changes in "Exponential" Data

Y_t	Y_{t-1}	$Y_t - Y_{t-1}$	$\dfrac{Y_t - Y_{t-1}}{Y_t}$
1987	1986	1000	0.0476190476
1988	1987	2900	0.1213389121
1989	1988	4100	0.1464285714
1990	1989	5000	0.1515151515
			sum = 0.4669016826

Source: Data developed for examples.

Notes: When using the exponential model, rounding numbers can produce large errors. It is best to use all of the digits produced by your calculator or computer in using this model. The formula for calculating the rate of change (r) may not be as accurate as using a log-linear growth rate, but provides reasonable accuracy and ease of use.

Example 2

An exponential model is used to project the historical data labelled "exponential" in Table 1. The calculations needed for this model are presented in Table 3.

Using the exponential model to project the historical data to 1993 yields:

$$Y_{t+n} = Y_t(1+r)^n,$$

$$\text{Subject}_{1993} = \text{Subject}_{1990}(1+r)^3,$$

$$= 33{,}000(1+r)^3.$$

The value for r is calculated from:

$$r = \frac{1}{m} \sum \frac{Y_t - Y_{t-1}}{Y_{t-1}}$$

$$= \frac{1}{3}(0.4669016826) = 0.1556338942.$$

$$\text{Subject}_{1993} = \text{Subject}_{1990}(1+0.1556338942)^3,$$

$$= 33{,}000\,(1.54333716) = 50{,}930.$$

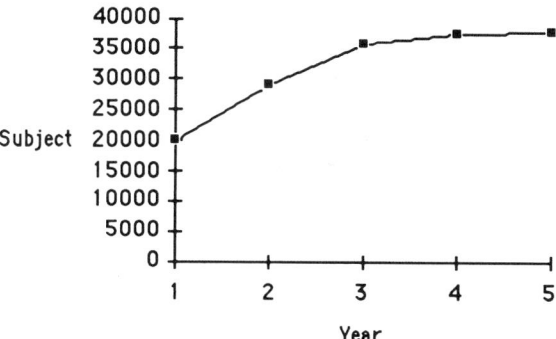

Figure 4. Time-series Plot of Modified Exponential Data. *Source*: Data developed for examples.

The exponential model produces a projection of 50,930 for 1993. This is a large increase from the 33,000 in 1990. Projections further into the future quickly become extremely large, which should alert the analyst that this model can rarely be used for long range projections.

■

Modified Exponential Projection Model

The *modified exponential projection model* may be appropriate where the rate of increase slowly declines as the subject being projected approaches some upper constraint on growth. This model is often used to project population for municipalities. The assumption is that as the town becomes completely developed (little vacant land for future growth), population growth will slow and become constrained. A time-series plot of the data labelled "modified exponential" in Table 1 is presented in Figure 4.

The mathematical formula for the modified exponential model is:

$$Y_{t+n} = C - [(C - Y_t)(p)^n]$$

$$p = \frac{1}{m} \sum \frac{C - Y_t}{C - Y_{t-1}},$$

where

Y = the measurement or activity being projected,
t = the unit of time of the data,
Y_t = the value of Y in the most recent time interval of the historical data and the starting point for the projection,
n = the number of units of time (in months, years, etc.),
Y_{t+n} = the activity n units of time from t,
C = the capacity or upper constraint on the projection
p = the portion of the unused capacity,
Y_{t-1} = the level of Y one time period before Y_t, and
m = the number of historical intervals over which the average annual growth is calculated.

Example 3

A *modified exponential model* is used to project the historical data labeled "modified exponential" in Table 1. In this example we are assuming that the capacity constraint (C) is 40,000. The calculations are presented in Table 4.

Table 4

Calculations to Determine the Portion of Unused Capacity (p)

YEAR	Y	C	$C - Y_t$	$C - Y_{t-1}$	$\dfrac{C - Y_t}{C - Y_{t-1}}$
1986	20,000	40,000			
1987	29,000	40,000	11,000	20,000	0.550
1988	36,000	40,000	4,000	11,000	0.364
1989	37,500	40,000	2,500	4,000	0.625
1990	38,000	40,000	2,000	2,500	0.800
					sum = 2.339

Source: Data developed for examples.

Using the modified exponential model to project the historical data to 1993 yields:

$$P = \frac{1}{4} 2.33900 = 0.5875,$$

$$\text{Subject}_{1993} = 40{,}000 - [(40{,}000 - \text{Subject}_{1990})(p)^3],$$

$$= 40{,}000 - [(40{,}000 - 38{,}000)(0.58475)^3],$$

$$= 39{,}600.$$

The modified exponential model produces a projection of 39,600 for this subject in 1993. A three year projection has approached the upper constraint on growth and there will be little increase in growth further into the future as the subject approaches the constraint.

Please note that a quadratic model is more complicated but may also be appropriate for fitting and or projecting this sort of data. ■

Ratio Projection Models

Ratio projection models are used when you decide to base your projection on an already prepared projection for some other entity. Usually this occurs when you are trying to make a projection for a smaller entity and a projection is available for a larger entity of which your entity is a part. For example, you want to project employment for a county and you find a projection for employment for the entire state. Ratio projection models allow you to project employment in the county as some ratio of employment in the state. Larger political units often have advantages that allow them to produce good projections. They generally have better data upon which to base projections and they often have larger staffs available to make projections on a regular basis.

The basic assumption of ratio projection models is that the past ratio of activity in your unit to that in some larger unit will continue into the future. For example, if your municipality has in the past contained 25 percent of the county's population, then we can use a ratio projection model that assumes your municipality will have 25 percent of the county's population a few years from now. This most simple of ratio projection models is sometimes called a *step-down* model or a *constant share* model. A variety of modifications to this basic model are possible.

$$Y^S_{t+n} = \frac{Y^S_t}{Y^L_t} Y^L_{t+n}$$

where

Y = the measurement or activity being projected,
s = the smaller unit for which we are making a projection,
t = the time period from which we are projecting,
n = the number of time periods into the future for which we are making a projection,
L = the larger unit for which we have a projection,
Y^S_{t+n} = activity being projected, in the smaller unit, in the time period n years from the present,
Y^S_t = activity to be projected, in the smaller unit, in the present,
Y^L_t = activity be projected, in the larger unit, in the present, and
Y^L_{t+n} = activity be projected, in the larger unit, n time units from the present.

Example 4

We want to estimate manufacturing employment for Milwaukee County in 1990. The manufacturing employment in Milwaukee County in 1985 was 115,067. The manufacturing employment in the State of Wisconsin in 1985 was 497,833. We have available a projection for manufacturing employment of 503,745 for the State of Wisconsin in 1990 in which we have some confidence.

$$\text{Milw manu emp}_{1990} = \frac{\text{Milw manu emp}_{1985}}{\text{Wisc manu emp}_{1985}} \text{Wisc manu emp}_{1990}$$

$$= \frac{115,067}{497,833} \times 503,745 = 116,433.$$

The ratio projection model uses the 1990 employment projection for Wisconsin to calculate the 1990 employment projection for Milwaukee County. The projection for Milwaukee County using 1985 data is that there will be 116,433 employees in the manufacturing sector in 1990. ∎

Further Readings

Ascher, William, *Forecasting: An Appraisal for Policy Makers and Planners*. Baltimore: Johns Hopkins University Press, 1978.

Greenberg, Michael R., Donald A. Krueckeberg, and Conny O. Michaelson, *Local Population and Employment Projection Techniques*. New Brunswick, New Jersey: The Center for Urban Policy Research, 1978.

Krueckeberg, Donald A., and Arthur L. Silvers, *Urban Planning Analysis: Methods and Models*. New York: John Wiley and Sons, 1974.

Pittenger, Donald B., *Projecting State and Local Populations*. Cambridge, Massachusetts: Ballinger Publishing Co., 1976.

U.S. Bureau of the Census, *Handbook of Statistical Methods for Demographers*, by A. J. Jaffe. Washington, DC: U.S. Government Printing Office, 1951.

Chapter 8
ANNUALIZING RATES OF CHANGE

Definition

Rates of change are important for describing data, whether the data relate to demographic, economic or social issues. Often we need to know the annual rate of change between two points several years or decades apart. Because of the *principle of compounding,* the annual rate is not merely the total rate divided by the number of years between the two points. Conversely, the total rate between two points is not the annual rate multiplied by the number of years between the two points.

Method

Annualizing data requires that we find the annual rate of change that, when compounded over a series of years, will equal the rate obtained when we divide the ending year by the beginning year, assuming a constant rate of change. The rate could be either increasing or decreasing over the time period.

Compounding of rates is easier to understand because of its use in banking, and our intuitive understanding of the increase to an investment over time. But the rate could be declining over time. Again, the compounded rate over time is not simply the annual rate multiplied by the number of years of the time period. First the principle of compounding, both increasing and decreasing, will be presented, then the principle of annualizing will be presented, for both increasing and decreasing rates.

Compounding Rates (Increasing)

$$\text{Total Rate (TR)} = (1 + r)^n - 1$$

where

r = annual rate of increase (will be a positive rate), and
n = number of years in the time period.

To convert the rate to a percentage multiply by 100.

Example 1a

Assume the annual rate is .05 and we wish to know the total increase over three years:

$$\text{TR} = (1 + r)^n - 1$$
$$= (1 + .05)^3 - 1$$
$$= 1.05^3 - 1$$
$$= .1576 \text{ or } 15.76\%.$$

The total rate of increase over the three years is .1576. This is equivalent to a 15.76% increase.

Check by compounding year by year at 1.05 (e.g., multiply each year's rate by 1.05 to obtain the next year's rate):

year 0 = 1.00,
year 1 = 1.05,
year 2 = 1.1025,
year 3 = 1.1576.

∎

Annualizing Rates of Change 91

Example 1b

Our Town bus ridership has been growing at 2.5% per year. Ridership in 1991 was 220,000 persons. If ridership keeps growing at 2.5% per year what will be the ridership in 1998? (Note that 2.5% converts to a rate of .025 by dividing 2.5 by 100.)

$$TR = (1 + r)^n - 1$$
$$= (1 + .025)^7 - 1$$
$$= 1.025^7 - 1$$
$$= .1887 \text{ or } 18.87\%.$$

Note: Projected value = (Total number at present) × (1 + rate), so the estimate for this example is computed as:

Ridership estimate for 1998:

Projected ridership = 220,000 × (1 + (.1887)) = 261,514.

■

Compounding Rates (Decreasing)

$$TR = (1 + r)^n - 1,$$

where

r = the annual rate of decline (will be a negative rate), and
n = number of years in the time period.

To convert the rate to a percentage multiply by 100.

Note that the formulas are the same for both the increasing and declining cases. The rate of change is positive for the increasing case and negative for the decreasing case. ■

Example 2a

The annual change is −5% and we wish to know the total decline over three years. (Note that −5% is equivalent to a rate of −.05.)

$$TR = (1+r)^n - 1$$
$$= (1 + (-.05))^3 - 1$$
$$= .95^3 - 1$$
$$= .8574 - 1$$
$$= -.1426 \text{ or } -14.26\%.$$

The total change over the three year period was $-.1426$ or -14.26%. Check by compounding year by year at $-.05$:

year 0 = 1.0000,
year 1 = .9500,
year 2 = .9025,
year 3 = .8574 (or a decline of .1426 or 14.26%). ∎

Example 2b

Your Town bus ridership is declining at 1.2% per year. Your ridership is now 360,000 persons per year. If your ridership keeps declining at 1.2% per year what will be the ridership in seven years?

$$TR = (1+r)^n - 1$$
$$= (1 + (-.012))^7 - 1$$
$$= .988^7 - 1$$
$$= .9190 - 1$$
$$= -.0810 \text{ or } -8.10\%.$$

Ridership estimate for year seven:

Projected ridership = $360{,}000 \times (1 + (-.0810)) = 330{,}840$. ∎

Annualizing Rates (Increasing)

The annual rate of change can be computed when we have data for the beginning and ending years of a period or when we have a rate for the

Annualizing Rates of Change

entire period. The following formula is used when we have beginning and ending data.

$$\text{Annual Rate (AR)} = \sqrt[n]{(B/A)} - 1,$$

where

A = beginning year,
B = ending year, and
n = number of years in the time period.

To convert to a percentage multiply by 100.

Note that you first find the nth root of B/A and then subtract 1 from it. It is very difficult to calculate n^{th} roots by hand, so we suggest that you purchase an inexpensive calculator that will allow you to find n^{th} roots. Look for a calculator with a $\sqrt[x]{y}$ function or a y^x key (the x in y^x must be the reciprocal of n, that is, the fifth root of y is $y^{1/5}$.) The instruction booklet that comes with the calculator will explain the step-by-step process for finding n^{th} roots. It is also possible to find an approximate n^{th} root by trial and error, that is, guess what the nth root might be and raise the number to the nth power and compare the answer to the number for which you wanted to find the nth root. Keep guessing the nth root and raising it to the nth power until you get a close approximation.

When we know the rate of change for the total time period, we use the following formula:

$$AR = \sqrt[n]{(1 + r)} - 1,$$

where

r = the rate of change for the total time period, and
n = number of years in the time period.

Note: We are finding the nth root of $1 + r$.

To convert to a percentage multiply by 100.

Example 3a

The value at the beginning of a three year period was 2. At the end of the period it was 3, or a 50% increase. We wish to know the annual rate. Using the beginning and ending values:

$$AR = \sqrt[n]{(B/A)} - 1$$
$$= \sqrt[3]{(3/2)} - 1$$
$$= 1.1447 - 1$$
$$= .1447 \text{ or } 14.47\%.$$

The annualized rate of increase over the three year period was .1447. This is equivalent to a 14.47% increase.

Using the rate of change:

$$AR = \sqrt[n]{(1+r)} - 1,$$
$$= \sqrt[3]{(1+.5)} - 1$$
$$= \sqrt[3]{(1.5)} - 1$$
$$= 1.1447 - 1$$
$$= .1447 \text{ or } 14.47\%.$$

Check by compounding at 1.1447:

year 0 = 2.0000,
year 1 = 2.2894,
year 2 = 2.6207,
year 3 = 3.0000.

∎

Example 3b

The population of Our Town was 62,000 on January 1, 1982. On January 1, 1990, it was 75,000. The total change between 1982 and 1990 was 20.97%. What was the annual rate of change?

$$AR = \sqrt[n]{(B/A)} - 1$$
$$= \sqrt[8]{(75,000/62,000)} - 1$$
$$= \sqrt[8]{(1.20968)} - 1$$
$$= 1.0241 - 1$$
$$= .0241 \text{ or } 2.41\%.$$

The annual population growth over the eight year period was 2.41%.

Annualizing Rates (Decreasing)

The formula can be applied when we have data for the beginning and ending years of a period or a rate for the entire period.

Using the formula for the beginning and ending years:

$$AR = \sqrt[n]{(B/A)} - 1$$

where

A = beginning year,
B = ending year, and
n = number of years in the time period.

Note: We are finding the nth root of B/A.
To convert to a percentage multiply by 100.

Using the formula for the rate for the entire period (note that this is the same formula as that for the increasing case, but that the rate of change is negative):

$$AR = \sqrt[n]{(1 + r)} - 1,$$

where

r = the rate of decline for the entire time period, (will be a negative rate), and
n = number of years in the time period.

Note: We are finding the nth root of $1 + r$.
To convert to a percentage multiply by 100.

Example 4a

The value at the beginning of a three year period was 4. At the end of the period it was 2, or a 50% *decrease*. We wish to know the annual rate of decline. Using the beginning and ending values:

$$\begin{aligned} AR &= \sqrt[n]{(B/A)} - 1 \\ &= \sqrt[3]{(2/4)} - 1 \\ &= .7937 - 1 \\ &= -.2063 \text{ or } -20.63\% . \end{aligned}$$

The annual change over the three year period was −20.63%.
Using the rate of change:

$$AR = \sqrt[n]{(1+r)} - 1,$$
$$= \sqrt[3]{(1+-.5)} - 1$$
$$= \sqrt[3]{(.5)} - 1$$
$$= .7937 - 1$$
$$= -.2063 \text{ or } -20.63\%.$$

Check by compounding year by year at −.2063:

$$\text{year } 0 = 4.0000,$$
$$\text{year } 1 = 3.1748,$$
$$\text{year } 2 = 2.5198,$$
$$\text{year } 3 = 2.0000.$$

∎

Example 4b

The population of Your Town was 85,000 on January 1, 1980. On January 1, 1988, it was 60,000. The total change between 1980 and 1988 was −29.41%. What was the annual change?

$$AR = \sqrt[n]{(1+r)} - 1,$$
$$= \sqrt[8]{(1+(-.2941))} - 1$$
$$= \sqrt[8]{(.7059)} - 1$$
$$= .9574 - 1$$
$$= -.0426 \text{ or } -4.26\%.$$

The annual change over the eight year period was −4.26%.

The point of this chapter was to demonstrate that the average rate of change for a time period is not simply the total rate divided by the number of years in the period. The error in this approach is simply too great to be accepted as a quick method. Rates of change must be annualized as shown.

Further Readings

Patton, Carl V., and David S. Sawicki, *Basic Methods of Policy Analysis and Planning*. Englewood Cliffs, New Jersey: Prentice-Hall, Inc., 1986.

Stokey, Edith, and Richard Zeckhauser, *A Primer for Policy Analysis*. New York: W. W. Norton and Co., 1978.

Texas Instruments, *BA•35 Quick Reference Guide*. Lubbock, Texas: Texas Instruments, Inc., 1989.

Chapter 9
SHIFT-SHARE ANALYSIS ....

Definition

Shift-share analysis is used to describe how the change in activity in one region is different from some reference region. It is usually used with economic data such as employment, value added, or regional product, and the reference region is often the nation (Dunn, 1960; Krueckeberg and Silvers, 1974). The technique can be used to separate the geographical components of temporal change in other contexts (Greenberg, 1980).

Shift-share analysis defines the *total shift* (sometimes called *net shift*) as the difference in output in the region being studied from what it would have been if it had grown at the same rate as total activity in the reference region. The total shift is equal to the sum of the proportionality shift and the differential shift. Each of these component shifts can be interpreted to provide insight into the change in activity level experienced by the region under investigation.

Method

The total shift (S_{is}), for a given activity i in a given location s, can be calculated directly using the formula:

$$S_{is} = \Delta E_{is} - \left(\frac{\Delta E_r}{E_{r,t-n}}\right) E_{is,t-n},$$

where:

Δ = change in activity between time period t and $t-n$,
E = the activity being measured (often employment),
i = the specific activity (the sector being studied),
s = the area or region being studied,
r = the reference region (often the nation),
$t-n$ = the initial time period of the study, and
$E_{is,t-n}$ = the activity being measured (E), in the sector of interest (i), in the region being studied (s), during the initial time period of the study ($t-n$).

The total shift is also equal to the sum of the proportionality shift and the differential shift:

$$S_{is} = P_{is} + D_{is}.$$

The *proportionality shift* is a shift in activity levels between two time periods measured for a specific activity. It is the difference between the reference region's (nation's) overall growth rate and the reference region's growth rate in the specific field being studied. Using employment data as an example, the proportionality shift is the difference between the growth rate for total employment and the growth rate for employment in a sector of interest, such as fabricated metals, multiplied by the starting period level of employment in fabricated metals in the region being studied.

The proportionality shift measures the "mix effect." It tells us if the specific field being studied (fabricated metals) is a fast growth field, a slow growth field, or a declining field relative to total growth for all sectors of the economy at the reference region (national) level. If the bracketed term in the following formula is positive, the specific field (fabricated metals) is a high growth field. A negative mix effect indicates that at the reference region level (nation) the specific activity grew slower than total activity.

$$P_{is} = \left(\frac{\Delta E_{ir}}{E_{ir,t-n}} - \frac{\Delta E_r}{E_{r,t-n}}\right) E_{is,t-n}.$$

The *differential shift* (D_{is}) is also a measure of change in a specific activity between two time periods. It calculates if the study region's growth rate for a specific activity (fabricated metals) grew more or less rapidly than the reference region's (nation's) growth rate for the specific activity. The differential shift measures the "competitive effect." It tells us about the relative competitive advantages of the region being studied. Using economic data, these competitive advantages might include distance to raw materials, distance to markets, transportation infrastructure, labor productivity, efficient production facilities, and other factors.

$$D_{is} = \left(\frac{\Delta E_{is}}{E_{is,t-n}} - \frac{\Delta E_{ir}}{E_{ir,t-n}} \right) E_{is,t-n}.$$

Example 1

A study was conducted using shift-share analysis of the change in employment in the fabricated metal products sector of the Wisconsin economy between 1980 and 1985. The results of the shift-share analysis are presented in Table 1. The fabricated metal products sector has a *Standard Industrial Classification* (SIC) code of 34.

Table 1

Shift-share Analysis of Fabricated Metal Products Employment.

Employment Changes in Fabricated Metals	Wisconsin	U.S.A.
1980 Fabricated Metals Products (SIC 34)	22,082	1,675,898
1980 Total Employment	1,493,482	74,835,525
1985 Fabricated Metals Employment	15,345	1,499,318
1985 Total Employment	1,624,756	81,119,257
Change in Fabr. Metals Employment	–6,737	–176,580
Change in Total Employment	131,274	6,283,732
Total Shift	–8,591	
Proportionality Shift	–4,181	
Differential Shift	–4,410	

Source: County Business Patterns.

The change in fabricated metal employment is calculated by subtracting the 1985 data from the 1980 data. For example, Wisconsin had employment of 22,082 − 15,345, or a change of −6,737.

$$\text{The proportionality shift} = \left(\frac{-176{,}580}{1{,}675{,}898} - \frac{6{,}283{,}732}{74{,}835{,}525}\right) 22{,}082 = -4{,}181.$$

$$\text{The differential shift} = \left(\frac{-6{,}737}{22{,}082} - \frac{-176{,}580}{1{,}675{,}898}\right) 22{,}082 = -4{,}410.$$

$$\text{The total shift} = (-4{,}181) + (-4{,}410) = -8{,}591.$$

In the period 1980 to 1985, the Wisconsin economy lost 6,737 employees (30.5%) in the fabricated metal products sector (SIC 34). During this period the economy of the country lost 10.5% of its employees in the fabricated metals product sector. Thus, while the country as a whole was losing jobs in the fabricated metal products sector, the state of Wisconsin was experiencing almost three times as great a loss of jobs in this sector in percentage terms.

The total shift measures the change in growth in the fabricated metals sector from what would have been experienced if this sector in Wisconsin had grown at the same rate as total employment in the country. Total employment in the country increased by 6,283,732 (8.4%) between 1980 and 1985. If fabricated metal products employment in Wisconsin had grown at that rate it would have been 8,591 (the total shift) greater in 1985 than it actually was.

The proportionality shift measures the "mix effect." It explains how much of the total shift in employment in fabricated metal products results from this sector losing jobs everywhere in the country. In Wisconsin between 1980 and 1985, 4,181 of the lost jobs in this sector are attributed to the fact the the entire economy of the United States was losing jobs in this sector. Wisconsin experienced this portion of its total job loss because the industries that comprise the fabricated metal products sector, taken as a whole, had declining employment in the country during this period.

The differential shift measures the "competitive effect." More than half of the total shift in the fabricated metal products sector in Wisconsin between 1980 and 1985 is the result of the competitive effect. There were competitive factors in Wisconsin during this period that resulted in the loss of 8,591 jobs in this sector. These jobs were lost because of competitive factors in Wisconsin that were less favorable to employment in fabricated metal products than conditions in the country as a whole. ■

Further Readings

Chapin, Francis Stuart, and Edward Kaiser, *Urban Land Use Planning*, Third Edition. Chicago: University of Illinois Press, 1979.

Dunn, Edgar, "A Statistical Analytical Technique for Regional Analysis," *Papers of the Regional Science Association.* **6**: 97–112, 1960.

Greenberg, Michael R., "A Method to Separate the Geographical Components of Temporal Change in Cancer Mortality Rates," *Carcinogenesis.* **1**: 553–557, 1980.

Krueckeberg, Donald A., and Arthur L. Silvers, *Urban Planning Analysis: Methods and Models.* New York: John Wiley and Sons, 1974.

Landis, John, "Electronic Spreadsheets in Planning, The Case of Shiftshare Analysis," *JAPA.* **51**(2): 216–224, 1985.

U.S. Department of Commerce, Bureau of the Census, *County Business Patterns, for the Years* 1980 *and* 1985 *for Wisconsin and for the United States.* Washington, DC: Government Printing Office.

Chapter 10
MULTIPLIER ANALYSIS ....

Definition

A *multiplier*, as the term is used in this chapter, is any constant term that is used in an arithmetic operation to estimate, apportion, or project some known quantity. Three uses of multipliers will be reviewed in this chapter: (1) the use of "standards" as multipliers to estimate requirements, (2) the use of a multiplier to make dollar values from different time periods comparable using the Consumer Price Index, and (3) the use of economic multipliers to project economic impacts. Each of these three types of multipliers will be treated separately.

Method

Published Standards

The first type of multiplier is the *published standard* that is used to estimate requirements. Standards are widely used in many fields, sometimes as final rules and sometimes only as rules of thumb that may later be modified. Examples might be the number of hospital beds needed for a given population, the number of acres of woodland needed to support a certain species of wildlife, or the percentage of

105

open space required in a large commercial development under a certain zoning classification. This type of multiplier can be used in any arithmetic operation.

Example 1

Employees with a firm have a standard office size of 200 net square feet. This standard can be used as a multiplier to either determine the floor area needed to contain offices for 15 employees (15 × 200 = 3000 net square feet) or to determine how many offices can be placed in a floor area of 4000 net square feet (4000/200 = 20 offices).

Another example of standards used as multipliers is parking standards. Standards tell us that parallel parking takes up 22 feet of curb per car and permits 4.5 cars parked per 100 feet of curb, while parking angled at 45 degrees to the curb occupies 11.3 feet of curb space per car and allows 8.2 cars per 100 feet of curb (De Chiara and Koppelman, 1969). These standards can be used as multipliers to determine how many feet of curb are needed to parallel park 100 cars:

$$100 \times 22 = 2200 \text{ feet of curb.}$$

A standard can be used as a multiplier to find how many cars can be parked along a 40-foot continuous length of curb using 45 degree angled parking:

$$40 \times 8.2/100 = 3.28, \text{ three cars.}$$

∎

Cost of Living Estimates

The second multiplier method is the use of a multiplier to make dollar values from different time periods or cost of living estimates for different locations comparable using the *Consumer Price Index* (CPI). Because of inflation, dollar values from different time periods must be adjusted before they can be directly compared. The most common form of the CPI presently in use is based on prices in the period 1982–1984. Previously, it was based on 1967 dollars. Always check the base period of the CPI you are using because some CPI's are calculated using different base periods. The CPI for each month in a given year tells us the inflation-adjusted value of 100 base period dollars in that

Multiplier Analysis

month of the given year in which we are interested and for a variety of locations. The CPI is a ratio of current prices to base period prices. It can be converted to a percentage by multiplying by 100.

The CPI is calculated for numerous classes of expenditures and for different regions and local areas. The CPI is sometimes referred to as the *CPI-U*, where the *U* indicates that it is an urban index thought to be applicable for about 80 percent of the U.S. population. The CPI is calculated from the prices of 400 items in a typical "market basket" of goods and services (Horwitz and Ferleger, 1980). The CPI is a weighted average in which the weights for each of the 400 items are the percentages of total income spent on each item. The weights thus sum to 100 percent (see Chapter 15, Indices).

An index is available for the major categories of apparel commodities, food and beverages, housing, fuel, household furnishings, medical care, services, transportation, apparel upkeep, and entertainment (Bureau of Labor Statistics, 1988). There are subcategories for each of these headings. For instance, under the heading "apparel commodities" there are individual CPI's for the subheadings: men's and boys', women's and girls', infants' and toddlers', foot wear, and other. CPI values are also available specific to regions and metropolitan sizes within each region, as well as for selected major urban areas (see Appendix 23).

There have been some changes in the CPI over the years. The most important has been a change in the treatment of housing costs as a result of their rapid increase relative to other costs. In the late 1960s and early 1970s, an experimental version of the CPI, known as the *CPI-U X1*, was available that has since become the standard CPI (Bureau of Labor Statistics, 1967). Those who use the CPI to adjust housing costs should adjust the CPI from earlier years to what was then the CPI-U X1 to be consistent with the present CPI.

If the CPI for September 1988 was 358.9, that tells us that $100 in September 1967 was comparable in value to $358.90 in September 1988 (Bureau of Labor Statistics, 1988). The consumer price index (CPI) can be used as a multiplier to convert dollars in one period to the real value (adjusted for inflation) of those dollars in another time period. Use of the CPI multiplier is necessary to compare the real value of dollars in different time periods.

Example 2

The CPI multiplier can be used to convert dollar values from earlier years to their present inflation-adjusted value. To find the value for all urban consumers of 1,000 1967 dollars in August 1990, we can use the CPI value as a multiplier (see Appendix 23). We solve the equation

$$\$1,000/100 = X/394.1,$$

where X = the value in 1990. This can be converted algebraically to

$$X = 394.1 (\$1,000)/100 = \$3,941.$$

The answer is that the 1,000 1967 dollars are equivalent in value to $3,941 in August 1990.

To find the August 1990 value of 1,000 September 1980 dollars, we must solve the simple ratio equation using CPI values for both September 1980 and August 1990. The September 1980 CPI for all items for all urban consumers is 251.7 (Bureau of Labor Statistics, 1980). The CPI for August 1990 is 394.1 (see Appendix 23). We solve the following equation for X, where X = the value of 1,000 1980 dollars in 1990:

$$1,000/251.7 = X/394.1.$$

This can be converted algebraically to

$$X = 394.1(\$1,000)/251.7 = \$1,565.75.$$

The $1,000 in September 1980 is equivalent to $1,565.75 in August 1990 when we take into account the effect of inflation.

This same procedure can be used to convert 1990 dollars to their 1975 value using the CPI values for 1990 and 1975 as multipliers. The September 1975 overall CPI for all urban consumers is 163.6 (Bureau of Labor Statistics, 1975). To convert the dollar values and correct for the effects of inflation, we solve the simple ratio equation for X, where X = the value of 1,000 August 1990 dollars in September 1975:

$$\$1,000/394.1 = X/163.6.$$

This equation can be converted algebraically to

$$X = \$1,000(163.6)/394.1 = \$415.12.$$

This tells us that $1,000 in August 1990 is equivalent in value to $415.12 in September 1975, when we adjust to remove the effects of inflation.

The CPI can be used as a multiplier to compare living costs in different parts of the country. A CPI is provided for both all urban consumers and for urban wage earners for selected urban areas each month. A family living in St. Louis that earns $3,500 per month can estimate the earnings needed in Boston to maintain the same standard of living.

To convert the dollar values from the cost of living in one city to those in a different city and correct for the effects of inflation, one must first find the CPI for all urban consumers for both cities. Using the *Monthly Labor Review*, one finds that the CPI for St. Louis was 128.0 and for Boston it was 138.0 (see Appendix 23, p. 269). One then solves the ratio equation for X, where X = the value of $3,500 St. Louis dollars in July 1990 that a family would need to maintain the same standard of living in Boston:

$$\$3,500/128.0 = X/138.0.$$

This equation can be converted algebraically to

$$X = \$3,500(138)/128 = \$3,773.44.$$

In order for the family from St. Louis to maintain their same standard of living in Boston, the family would need to earn $3,773.44 per month. ■

Economic Multipliers

The third type of multiplier discussed in this chapter is the *economic multiplier*. A variety of techniques have been developed to estimate multipliers for regional income or regional employment (Isard, 1960). Once these multipliers have been estimated or calculated and some assumptions have been made about their reliability in the future, these economic multipliers are used to project income or employment in the future. The earliest and still most widely used methods of developing economic multipliers are the Economic Base Study and Input–Output Analysis (Tiebout, 1962; Miernyk, 1965; Shah, 1979). Both of these methods use sample data to calculate multipliers specific to selected industries. Both techniques assume that exports (the basic sector) drive the regional economy by bringing in money from outside the region. The multipliers are used to estimate the impact on the regional economy of new economic activity, usually additional export earnings.

The use of economic multipliers assumes that one dollar of additional export earnings will have greater impact on the regional economy than one dollar of local earnings. In theory the additional dollar of export earnings starts an infinite number of expenditure cycles in the economy. Some portion of the income earned from the exports is spent on consumption during each of these cycles. When the money is spent on consumption in the region, business people in the region earn additional income and spend some portion of their additional income. This recycling of some portion of the original new money through the infinite cycles of the multiplier produces the multiplier effect. The result is that total income in the region is greater by some multiple of the original export earnings to the region.

Economic multiplier analysis uses these assumptions and the calculated or estimated multipliers to project the total impact on the regional economy of new money coming into the region. The multipliers might be used to project the impact of an expansion to an existing firm in your region that produces lawnmowers, or to project the impact of a firm from outside your region that has decided to relocate in your region. In both cases, economic multipliers specific to the industry being considered are needed. Sometimes, multipliers specific to the four digit SIC code (standard industrial classification) level are available, and sometimes less specific multipliers are all that are available (U.S. Department of Commerce, 1972). It is best to use multipliers as specific to the industry in question as possible and to use multipliers determined for your region. Sometimes, multipliers specific to your region are not available and multipliers to a larger region or even national multipliers are used (Ritz, 1980).

Despite several important problems, economic multiplier (impact) analysis is widely used. The greatest problem in their use is that the calculation of economic multipliers by means of input–output analysis or an economic base study is a complex and expensive process. There are also several conceptual problems, for instance: (1) the multipliers can not take into account technological advances made after the multipliers were calculated but in use in the economy at the time the economic impact study is conducted, (2) the models used to calculate multipliers assume that resources used must be proportional to output, implying that changes in production levels ignore economies or diseconomies of scale in production, and (3) the multipliers are assumed to have been calculated and used during periods of maximum production; this is not always true and was a problem with the multipliers produced from the 1958 national input–output table. Despite these problems, economic multipliers are calculated and then widely used to estimate impacts.

Example 3

We want to project the impact on the economy of the Milwaukee region of a contract signed by the Harnischfeger Corporation, a local firm, to sell one million dollars of construction equipment to a foreign company. Multiplier analysis can be used to estimate the effect of increased sales of construction equipment. In Milwaukee, Wisconsin, the multiplier for the manufacture of construction machinery (SIC 3531) was calculated in a recent study to be 3.37 (note that multipliers are typically larger in major metropolitan areas than in small cities). The multiplier times the machinery sales value produces a figure of $3.37 million in total value to the Milwaukee economy because of the sale of one million dollars of additional construction equipment by the Harnischfeger Corporation. ∎

Further Readings

Bureau of Labor Statistics, "Three Standards of Living for an Urban Family of Four Persons," *Bulletin No. 15705.* Spring, 1967.

Bureau of Labor Statistics, *Monthly Labor Review*, Current Labor Statistics. **98**(12): 95–102, 1975.

Bureau of Labor Statistics, *Monthly Labor Review*, Current Labor Statistics. **103**(12): 97–104, 1980.

Bureau of Labor Statistics, *Monthly Labor Review*, Current Labor Statistics. **111**(11): 78–82, 1988.

Bureau of Labor Statistics, *Monthly Labor Review*, Current Labor Statistics. **112**(3): 86, 1989.

De Chiara, Joseph, and Lee Koppelman, *Planning Design Criteria.* New York: Van Nostrand Reinhold Co., 1969.

Horwitz, Lucy, and Lee Ferleger, *Statistics for Social Change*. Boston: South End Press, 1980.

Isard, Walter, *Methods of Regional Analysis*. Cambridge, Massachusetts: Technology Press of MIT and John Wiley and Sons, 1960.

Krueckeberg, Donald A., and Arthur L. Silvers, *Urban Planning Analysis: Methods and Models*. New York: John Wiley and Sons, 1974.

Miernyk, William, *The Elements of Input–Output Analysis*. New York: Random House, 1965.

Ritz, Philip, *Definitions and Conventions of the 1972 Input–Output Study*. Washington, DC: U.S. Department of Commerce, Bureau of Economic Analysis, Staff Paper No. 80–034, 1980.

Shah, Praful, "Economic Base Studies: Elements, Forecasting, Fiscal Impact, Evaluation," pp. 599–613, *in* F. S. So, I. Stollman, F. Beal, and D.S. Arnold (eds.), *The Practice of Local Government Planning*. Washington, D.C.: International City Management Association, 1979.

Tiebout, Charles, *The Community Economic Base Study*. New York: Committee for Economic Development, 1962.

U.S. Department of Commerce, *Standard Industry Classification (SIC) Manual*. Washington, DC: Government Printing Office, 1972.

PART 4
How to Obtain and Validate Data

Chapter 11
SAMPLING

Definition

Data needed for decisions come from a variety of sources. Sometimes data collected by others, by the U.S. Bureau of the Census, for example, are available and meet our needs. Sometimes other types of secondary data are available that address the problem. But often the data needed to examine an issue are not at hand or available from others, and we need to collect original data. Because time is usually short, we often collect these data through sample surveys. *Statistical sampling*, then, is the process by which a portion of the whole (population or universe) is selected for examination with the intent to generalize from that sample to the entire population.

Samples are selected because data obtained in this way can be collected quickly when needed and at a lower cost than collecting data on every element of the subject under study. The quality of information collected in this way can even be more accurate than the data that might be collected from a 100% survey, since, for the budget available, more attention can be paid to the sample, more in-depth data can be collected, missing data or errors can be rectified, and a follow-up can be conducted. Moreover, even if money were no object, we likely would not conduct an enumeration or 100% census, since we need only a certain level of accuracy for most projects. The level of accuracy required, that is, whether we need only a close approximation or must obtain a precise number, will affect the type and size of sample taken. (How large a sample to take is discussed in Chapter 13, Sample Size.) The key concepts underlying sampling are those of *randomness* and *representativeness*. All members of the group being surveyed must have a known chance of being selected in the sample. Note that haphazardly asking people who pass by a street corner their opinion on a presidential candidate would not be a random sample. The random sample must be conducted in a scientific, unbiased manner to assure that all people in the population under study have an equal chance of being selected.

Method

Samples can be selected in several ways that, when done properly, assure randomness. These methods essentially assure that the persons selected in a sample represent the full range and characteristics of the people in the group from which the sample is selected, and that each person's chance of being selected for the sample is independent of any other person's chance of being selected. We will discuss simple random sampling, systematic sampling, stratified sampling, cluster sampling, and disproportionate sampling. We will also discuss the process of weighting sample results to adjust for disproportionate sampling and varying response rates, and to produce population estimates.

Simple Random Sampling

This approach involves selecting elements or members from a group or population at random. Usually the elements are listed in a convenient order (say alphabetically) but this is not required, and a random number table is used to select a given number of elements from the list.

Example 1

We wish to select a 25% sample of persons in a drug rehabilitation program. The 20 persons in the program were each assigned a number as they enrolled in the program, and that list is given in Table 1. A 25% sample of 20 persons is 5 persons. To select these 5 persons we use a random number table to identify the ID number of the persons to select.

A typical random number table is reprinted in part as Table 10 at the end of this chapter. A portion of that table is adapted as Table 2 in order to illustrate how to use such a table. Because we will need only two-digit numbers, the table has been reformatted into columns of two-digit numbers. In using a random number table, we first select a random start in the table and then select numbers from that point onward.

Table 1
Rehabilitation Participants

1	Joe
2	Fred
3	Susan
4	Mary 1
5	Martha
6	Pete
7	Jack
8	Jeanne
9	Joan
10	Kathy
11	Mary 2
12	Sam
13	Geoff
14	Barbara
15	Ellie
16	Andy
17	Eric
18	George
19	Sarah
20	John

Table 2
Random Number Table Excerpt

10	09	73	25	33
37	54	20	48	05
08	42	26	89	53
99	01	80	25	29
12	80	79	99	70

Source: The RAND Corporation. *A Million Random Digits with* 100,000 *Normal Deviates*, New York: The Free Press, 1955, p. 1.

For a 25% sample select 10, 8, 12, 9, and 1.

Assume for this example that we opened the random number table to a random page and began selecting random numbers with the first one on the page. Since the people on our list are numbered from 1 to 20, we select the first five numbers in the table that fall between 1 and 20. In this case we would begin with 10, then 8, then 12, etc. When selecting a large number of cases, the random numbers may repeat ones we have already selected. When this happens, we simply continue selecting numbers that fall within the range until we have the proper number of unduplicated cases. ∎

Systematic Sampling

This method is used for selecting elements from a list. The members of the group are listed in some type of actual or implied order (alphabetically, by street address, by ID number, etc.), and a given number of members is selected from the group by picking every nth item from the list, with the starting point selected randomly. The list, by the way, should not have its items listed in a cyclical pattern so that the sample would yield biased results (e.g., in a sample of apartments always selecting the manager's apartment).

Example 2

We wish to select a sample of persons from the residential listings in the Milwaukee phone book. (Other lists might include the voter rolls, the city directory, a membership roster.) There are 384,000 listings in the phone book (800 pages, 4 columns per page, 120 entries per column). We want a sample of 400 names and use the following process:

$$I = N/n,$$

where

I = interval,
N = population, and
n = sample size desired.

From the 384,000 entries in the Milwaukee phone book we desire a sample of 400 persons. Thus, $384,000/400 = I = 960$.

Using a random number table to determine the starting point between 1 and 960, we would select every 960th number to yield a systematic random sample of 400. Since this process requires the numbering of each entry, which would be tedious for a large list, we use the alternative procedure of selecting a page interval and then selecting an item from that page. For example,

We would thus:
$$\frac{800 \text{ pages}}{\text{sample of } 400} = 2.$$

- select a name from every second page,
- select the starting page randomly, using a random number table to select either page 1 or page 2 as the starting page (alternatively any page could be selected as the starting point if we circle back through the beginning of the book to return to the starting point),
- select a column randomly, using a random number table,
- select the line randomly, using a random number table, and
- use the same column and line for each page selected. ■

Stratified Sampling

Stratified sampling is a modification of the preceding methods that helps assure that we obtain a representative sample by breaking the population into homogeneous subsets that are then sampled. A population could, for example, be stratified by race, age, and sex, or other appropriate variables. A sample then may be taken from each strata based on its proportion of the total, or the strata may be listed in a continuous order and a systematic sample may be taken.

Example 3

We wish to sample university students by year in school. We use a sample stratified by year and then select a sample from each year in proportion to the size of the class. The four samples could be random

Table 3

A Stratified Sample of University Students

Freshmen	Sophomores	Juniors	Seniors
101	201	301	401*
102*	202	302	402
103	203*	303*	403
104	204	304	404
105	205	305	405
106	206	306	406*
107*	207	307	407
108	208*	308*	408
109	209	309	409
110	210	310	410
111	211	311	411*
112*	212	312	412
113	213*		413
114	214		414
	215		415
			416*
			417
			418
			419

Source: Analyst's example.

samples, each in proportion to the size of the class, or the four classes could be listed continuously, and a systematic sample could be taken. Using the second method, we select a 20% sample from this list of students in Table 3, each student being identified by his or her ID number. Since we have a total of 60 students, a 20% sample would yield 12 cases. The sampling interval would be 5 (i.e., 60/12). We determined student number 2 as the starting point by using a random number table to select a number between 1 and 5. The selected students are identified with asterisks. ■

Cluster Sampling

Cluster sampling is used when it would be difficult to compile a list of elements in the population, for example, of households in a census

Sampling

tract or households on all city blocks. In order to simplify the selection of cases, elements are clustered or grouped together, and then selected clusters are sampled. That is, city blocks are clustered together to create census tracts and households are clustered together as city blocks. Cluster sampling also reduces the cost of a sample survey when a large geographic area is involved, for example, when selecting households for personal interviews in Los Angeles or Chicago. First, a sample of city blocks would be randomly selected and then households on those blocks would be randomly selected for interviewing.

Example 4

Assume, for simplicity, that we live in a city of four blocks with a total of 120 households as shown in Table 4. We wish to select households on 50% of the blocks to interview about dish soap preferences. We therefore will select two blocks. On each block we will interview five households, a standard convention in order to interview enough households to have some confidence in the representativeness of the results. We use a cluster sampling procedure in which we first select 50% of the blocks and then select five households from each block. In order to give each household an equal chance of being selected, we have to give the blocks with larger numbers of households a greater chance of being selected as a sample block. We do this by basing the sample on the number of households in town rather than on the number of blocks, as follows:

Procedure:

1. Select clusters (in this case, blocks) on the basis of their size (that is, give the larger clusters a greater chance of being selected into the sample).
2. Select the same number of elements from each cluster (so that elements in larger clusters have a smaller chance of selection).

This dual procedure balances the dissimilar cluster and element probabilities so that each household has an equal chance of being selected.

Table 4

Selecting a Cluster Sample from a City of Four Blocks

Blocks (Clusters)	Households On Block	Cumulative Total Households	Random Numbers Related To Households
A	20	20	1–20
B	60	80	21–80
C	10	90	81–90
D	30	120	91–120

The goal is to select two blocks and then five households from each of the two blocks. We use the following procedure:

- determine the total number of households, in this case, 120,
- select two random numbers from 1–120 to identify the blocks, using an interval of 60 (120/2 = 60),
- take a random start between 1 and 60 (e.g., 17 selected from a random number table),
- select the blocks on which households 17 and 77 (17 + 60) are located (blocks A and B),
- select the households to be sampled (5 per block),
- determine the sampling intervals ($A = 20/5 = 4$), ($B = 60/5 = 12$), and
- on each block select the starting household from a random number table and use the interval to select the balance of households (e.g., for block B, assume the random start was 9, then the households selected would be 9, 21, 33, 45, 57).

Note that in this procedure, each household has an equal chance of being selected. The probability of being selected is the product of the probabilities of the block being selected and the household being selected from among households living on the block. For example:

Sampling

Probability of a household on block *A* being selected:

Probability of the block being selected = 20/120 = .166.
Probability of a household being selected from the block = 5/20 = .25.
Overall probability of a household being selected from block *A* = .166 × .25 = .0145 (rounded).

Probability of a household on block *B* being selected:

Probability of the block being selected = 60/120 = .50.
Probability of a household being selected from the block = 5/60 = .083.
Overall probability of a household being selected from block *B* = .50 × .083 = .0145 (rounded).

If you expected 20 households on the block and planned to sample 5 households, but in the field you found 40 households on the block, you should use the planned sampling interval (e.g., 4; thus selecting 40/4 = 10 households). This will maintain the original probability of each household being selected (e.g., 5/20 (planned) = 10/40 (actual) = .25).

■

Disproportionate Sampling

We use disproportionate sampling to ensure that a subpopulation contains enough cases for analysis. In order to obtain a minimum number of respondents from a particular strata, we may take a larger percentage sample from that strata (e.g., an area of the city with a small number of residents who have a particular characteristic we wish to study).

Example 5

We are interested in information about children who may be undernourished in our city. We suspect that this might be a particular problem in one area of the town. That area has a smaller total population

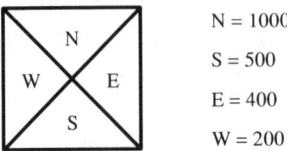

N = 1000
S = 500
E = 400
W = 200

Figure 1. The Distribution of Population in Our Town. *Source*: Analyst's example.

Table 5

Selection of Disproportionate Samples for Our Town

Area	Number of Households	Percentage Sample	Assumed Response Rate	Expected Yield
N	1000	10%	50%	50
S	500	20%	50%	50
E	400	25%	50%	50
W	200	50%	50%	50

Source: Analyst's example.

than the other areas of town, so we take a larger sample to assure that we will obtain enough sample cases to conduct a reliable analysis. The numbers of households in each section of the town are shown in Figure 1.

We wish to obtain data from approximately 50 households in each area. Since we expect a 50% response rate, we plan to contact 100 households in each area. We take disproportionate samples from the four areas as indicated in Table 5. ■

Weighting

It may be necessary to weight the results of a sample to adjust for a disproportionate sample, to correct for a misestimate of the size of a cluster or strata, to adjust for different response rates, or to make population estimates.

Example 6

After a disproportionate sample is collected, we might want to combine the data for all areas so we can talk about our findings for the city as a whole. To do this, we must weight the data by the inverse of the sampling ratio so responses from all areas have equal representation. The same thing must be done to correct for different response rates we may have gotten from various subareas. (We may wish to weight to adjust for different response rates even when we do not sample disproportionately in order to obtain population estimates.) This weighting process is shown below for the example in Table 5. Remember that we sampled disproportionately among the four areas in order to obtain samples of approximately equal minimum size.

Table 6 shows the percentage sampled in each subarea as well as the actual response from each area. That is, in the northern part of town we sampled 10% of the households and 80% of these households responded. We use these two pieces of information to compute a *weightfactor* for the responses from each area.

To create a weightfactor that will allow us to produce population estimates for the East area, where 25% of the people were sampled and 50% of these people responded, we find the product of the inverse of these two rates. That is, .25 of the households were sampled, so the inverse is 1/.25 = 4. Of the households sampled, .50 responded, so the inverse is 1/.50 = 2. Thus, the weight factor is 4 × 2 = 8. This means that if each response from the East area is multiplied by 8, the results are assumed to be those that we would have received had all house-

Table 6

Sample and Response Percentages for Our Town

Area	Percentage Sampled	Actual Response
N	10%	80%
S	20%	75%
E	25%	50%
W	50%	40%

Source: Analyst's example.

Table 7
Population Estimate Weightfactors for Our Town Sample

Area	Sample Inverse	Response Inverse	Computation	Weightfactor
North	1/.10	1/.80	10 × 1.25 =	12.5
South	1/.20	1/.75	5 × 1.333 =	6.665
East	1/.25	1/.50	4 × 2.0 =	8.0
West	1/.50	1/.40	2 × 2.5 =	5.0

holds been surveyed and all had responded. Table 7 shows these computations for all four areas of town. As another example, each response from the North side of town is counted as if it were given by 12.5 people. Each response from the West side is counted as 5.0 responses, etc.

Note that sometimes people add the individual weights (e.g., 10 + 1.25 = 11.25) but this is an error and will underestimate the weightfactor.

When we are working with numbers of persons sampled and number of respondents rather than percentages, the computations are made as follows:

$$\text{Weightfactor} = \frac{\text{total population}}{\text{number sampled}} \times \frac{\text{number sampled}}{\text{number responding}}.$$

For the North area discussed above, the computation becomes:

$$\text{Weightfactor} = \frac{1000}{100} \times \frac{100}{80} = 10 \times 1.25 = 12.5.$$

Or, more simply, when we know both the total population and the number responding, the following is the mathematical equivalent to the above:

$$\text{Weightfactor for N} = \frac{\text{total population}}{\text{number responding}} = \frac{1000}{80} = 12.5.$$

Prove to yourself that you will obtain the same answers as in Table 7 by computing the weightfactors in this manner. It is important to remember, however, that the weightfactor in this case is comprised of both the weight for the sampling ratio and the weight for response rate.

If the subarea data are never combined to create a description of the city as a whole, there would be no need to weight the responses. The percentages and values for each subarea would represent them accurately. Unfortunately, unweighted subarea data are sometimes combined by inexperienced analysts without the reader or audience being aware that the resulting total is inaccurate. ■

Data Quality

High quality data and sound data collection techniques are essential to good analysis. The sampling procedures presented above are only as good as is their implementation. If the sample is selected properly, but poor field methods are employed, or if a faulty data collection instrument or questionnaire is used, then not much confidence can be placed in the results. Similarly, if poor quality secondary data are used, then little confidence can be placed in the results of the analysis performed on those data as well. Major sources of secondary data are identified in Table 8, and questions to ask about the quality of those data are presented in Table 9. How, why, and by whom the data were collected are useful clues to its veracity. As a general rule, data that were systematically collected through a random sample as part of an ongoing process by trained staff under a skilled administrator are most likely to be valid.

Table 8

Data Source Checklist.

Libraries
Local libraries
University libraries
State libraries
Federal depository libraries
Agency libraries
Interlibrary loans

Federal agencies
Census Bureau and State Data User
 Centers
Department of Agriculture
Department of Commerce
Department of Health and Human
 Services
Department of Housing and Urban
 Development
Bureau of Labor Statistics
Internal Revenue Service
National Center for Health Statistics
Social Security Administration

State agencies
State planning agency
State departments of transportation,
 health, education, etc.
State budget bureau
State archive
State reference library
State license bureau (automobile
 and other)

Local agencies
Local and regional planning agencies
Departments of public works, building
 inspection, zoning, etc.
Community development departments

City and county assessors' offices
County extension agencies
Law enforcement agencies
Public health offices
Social service agencies
School districts
Housing authorities

Other public and quasi-public bodies
Water and sanitary districts
Gas and electric companies
Telephone companies
Transit districts
Park and recreation districts
Public health districts
Economic development districts

Survey research organizations
University-affiliated organizations
Private firms
Radio and television stations
Newspapers

Private organizations
Chamber of commerce
Boards of realtors
Voters' leagues
News media organizations
F. W. Dodge Division of McGraw-Hill
 Information Systems Company
Rand McNally
Sanborn Map Company
R. L. Polk Company
Donnelly Marketing Information
 Services
Conference board
Consulting firms

Source: Carl V. Patton, "Information for Planning," in *The Practice of Local Government Planning (Second Edition)*, Frank S. So and Judith Getzels, eds. Washington, D.C.: International City Management Association, 1988, p. 474.

Table 9
Data Quality Checklist.

How were the data collected?	**When were the data collected?**
Systematically/Haphazardly	After planning/During a crisis
Random sample/Nonrandom sample	Recently/In the past
Impartial third party/Program personnel	
	Who collected the data?
Why were the data collected?	Trained/Untrained personnel
Ongoing monitoring/Response to a crisis	Experienced/Inexperienced personnel
To respond to an internal need/To fulfill an external requirement	High-level/Low-level staff
	Highly/Not highly regarded staff
	Organized/Unorganized director
	Skilled/Unskilled communicator

Source: Carl V. Patton, "Information for Planning," in *The Practice of Local Government Planning (Second Edition)*, Frank S. So and Judith Getzels, eds. Washington, D.C.: International City Management Association, 1988, p. 475.

Further Readings

Babbie, Earl R., *Survey Research Methods*. Belmont, California: Wadsworth, 1973.

Kish, Leslie, *Survey Sampling*. New York: John Wiley, 1965.

Patton Carl V, "Information for Planning," in *The Practice of Local Government Planning* (Second Edition), Frank S. So and Judith Getzels, eds. Washington, D.C.: International City Management Association, 1988 pp. 472–499.

The RAND Corporation, *A Million Random Digits with 100,000 Normal Deviates*. New York: The Free Press, 1955.

Sudman, Seymour, *Applied Sampling*. New York: Academic Press, 1976.

Table 10
Random Number Table.

00000	10097	32533	76520	13586	34673	54876	80959	09117	39292	74945
00001	37542	04805	64894	74296	24805	24037	20636	10402	00822	91665
00002	08422	68953	19645	09303	23209	02560	15953	34764	35080	33606
00003	99019	02529	09376	70715	38311	31165	88676	74397	04436	27659
00004	12807	99970	80157	36147	64032	36653	98951	16877	12171	76833
00005	66065	74717	34072	76850	36697	36170	65813	39885	11199	29170
00006	31060	10805	45571	82406	35303	42614	86799	07439	23403	09732
00007	85269	77602	02051	65692	68665	74818	73053	85247	18623	88579
00008	63573	32135	05325	47048	90553	57548	28468	28709	83491	25624
00009	73796	45753	03529	64778	35808	34282	60935	20344	35273	88435
00010	98520	17767	14905	68607	22109	40558	60970	93433	50500	73998
00011	11805	05431	39808	27732	50725	68248	29405	24201	52775	67851
00012	83452	99634	06288	98083	13746	70078	18475	40610	68711	77817
00013	88685	40200	86507	58401	36766	67951	90364	76493	29609	11062
00014	99594	67348	87517	64969	91826	08928	93785	61368	23478	34113
00015	65481	17674	17468	50950	58047	76974	73039	57186	40218	16544
00016	80124	35635	17727	08015	45318	22374	21115	78253	14385	53763
00017	74350	99817	77402	77214	43236	00210	45521	64237	96286	02655
00018	69916	26803	66252	29148	36936	87203	76621	13990	94400	56418
00019	09893	20505	14225	68514	46427	56788	96297	78822	54382	14598
00020	91499	14523	68479	27686	46162	83554	94750	89923	37089	20048
00021	80336	94598	26940	36858	70297	34135	53140	33340	42050	82341
00022	44104	81949	85157	47954	32979	26575	57600	40881	22222	06413
00023	12550	73742	11100	02040	12860	74697	96644	89439	28707	25815
00024	63606	49329	16505	34484	40219	52563	43651	77082	07207	31790

Sampling

00025	61196	90446	26457	47774	51924	33729	65394	59593	42582	60527
00026	15474	45266	95270	79953	59367	83848	82396	10118	33211	59466
00027	94557	28573	67897	54387	54622	44431	91190	42592	92927	45973
00028	42481	16213	97344	08721	16868	48767	03071	12059	25701	46670
00029	23523	78317	73208	89837	68935	91416	26252	29663	05522	82562
00030	04493	52494	75246	33824	45862	51025	61962	79335	65337	12472
00031	00549	97654	64051	88159	96119	63896	54692	82391	23287	29529
00032	35963	15307	26898	09354	33351	35462	77974	50024	90103	39333
00033	59808	08391	45427	26842	83609	49700	13021	24892	78565	20106
00034	46058	85236	01390	92286	77281	44077	93910	83647	70617	42941
00035	32179	00597	87379	25241	05567	07007	86743	17157	85394	11838
00036	69234	61406	20117	45204	15956	60000	18743	92423	97118	96338
00037	19565	41430	01758	75379	40419	21585	66674	36806	84962	85207
00038	45155	14938	19476	07246	43667	94543	59047	90033	20826	69541
00039	94864	31994	36168	10851	34888	81553	01540	35456	05014	51176
00040	98086	24826	45240	28404	44999	08896	39094	73407	35441	31880
00041	33185	16232	41941	50949	89435	48581	88695	41994	37548	73043
00042	80951	00406	96382	70774	20151	23387	25016	25298	94624	61171
00043	79752	49140	71961	28296	69861	02591	74852	20539	00387	59579
00044	18633	32537	98145	06571	31010	24674	05455	61427	77938	91936
00045	74029	43902	77557	32270	97790	17119	52527	58021	80814	51748
00046	54178	45611	80993	37143	05335	12969	56127	19255	36040	90324
00047	11664	49883	52079	84827	59381	71539	09973	33440	88461	23356
00048	48324	77928	31249	64710	02295	36870	32307	57546	15020	09994
00049	69074	94138	87637	91976	35584	04401	10518	21615	01848	76938

Source: The RAND Corporation, *A Million Random Digits with 100,000 Normal Deviates*. New York: The Free Press, 1955, p. 1.

Chapter 12
CONFIDENCE LEVELS

Definition

When working with data obtained through a sample, we need to know the accuracy of the estimates we make based on those data. That is, we want to know how close the values estimated through our sample are to the values in the population from which the sample is taken. The size of sample, discussed in the next chapter, affects the extent to which we can place confidence in the data collected through a survey. But before the size of sample can be established, we need to determine the *level of accuracy (or confidence) required* for the problem we are addressing. The level of accuracy is measured through the concepts of *confidence level* and *confidence interval*.

In order to make these estimates, we will use the properties of the *normal curve*, which is a probability distribution that allows us to determine the relationship of any value in a sample of normally distributed variables to the mean value of the sample. As we will see later, this information will be useful in helping us determine how likely it is that certain values would be found in a sample. To analyze a set of data in this way, we set the mean of the sampling distribution equal to zero and then state all of the other values, both above and below the mean, in terms of their relationship to the mean in units of the standard deviation (Runyon and Haber, 1980, pp. 105–116). While this may sound involved, the following step-by-step example shows how and why it is done.

Method

In sampling, we want the values we estimate through the sample to be relatively close to the actual values in the population. We realize, however, that the estimated and actual values will seldom ever be identical. In sampling, our goal is to be roughly right within certain time and cost constraints. The question is, How often are we willing to be wrong in our estimates: 1 out of 100 times, 5 out of 100 times, 10 out of 100? This is the *confidence level*, often referred to as the significance level. The other part of the question is the *confidence interval*. In making an estimate, how far from the true value are we willing to be? For example, in an estimate of mean income, are we willing to accept a value plus or minus $1,000 of the true mean, or will we accept only a value plus or minus $5 of the true mean?

Here is an example of how the confidence level and confidence interval are related.

In estimating the true mean age of professional football players in the National Football League, we could make the following statements:

- We are 99% confident that the true mean age of the players in the NFL is between 15 and 60.
- We are 95% confident that the true mean age is between 20 and 40.
- We are 68% confident that the true mean age is between 25 and 30.

Note that as the confidence interval was *narrowed* we had to *lower* the confidence level.

In order to talk in statistical terms about confidence level and confidence interval, we need to use concepts related to the *normal curve*. We will not discuss the curve in great detail, just enough so you can use it to estimate properly sized samples.

Using the properties of the normal curve and the normal curve area table, we can answer a variety of questions: What is the probability that a value greater than x will occur? What is the probability that

Confidence Levels 135

values between x_a and x_b will occur? What is the probability that a value below x_c will occur? (Technical note: The probability of a single value of x occurring is zero since the area above a single value has a line width of zero.)

To use the normal curve and its related tables, we must standardize the data by converting them into z scores. The standardized values are obtained by subtracting the mean from the original value to get the deviation which is then divided by the standard deviation. (See Chapter 1 and the appendix to this chapter for a review of the standard deviation.) This procedure will allow us to measure in a standardized way (in the number of standard deviation units) how far any given value lies from the mean value. The following formula is used to produce the z score:

$$z_i = \frac{x_i - \mu}{\sigma},$$

where

z_i = the z score related to a particular value,
x_i = the particular value,
μ = the population mean, and
σ = the population standard deviation.

When we are working on a problem where the population's mean and standard deviation are unknown, we must substitute the sample's mean and standard deviation as an estimate of the population values in order to apply this formula. This is acceptable when the sample size is at least 30.

When the x values are converted to z values through the above formula, the z values will have a mean of zero and a standard deviation of 1 as shown in the z curve in Figure 1. Each z value is the number of standard deviations from the mean. For example, $z = 1.0$ is one standard deviation above the mean and $z = -2.5$ is two and one-half standard deviations below the mean.

The *normal curve table* presents information about the way data are distributed within the normal curve and allows us to make estimates about the relationship among values we determine through sampling (Table 1). The normal curve table relates z scores to the area between the mean of the distribution (which is zero) and the z score. For a given z score, the table reports the area from the mean of 0 to that z score. The figures given in the table are absolute values, and are for areas

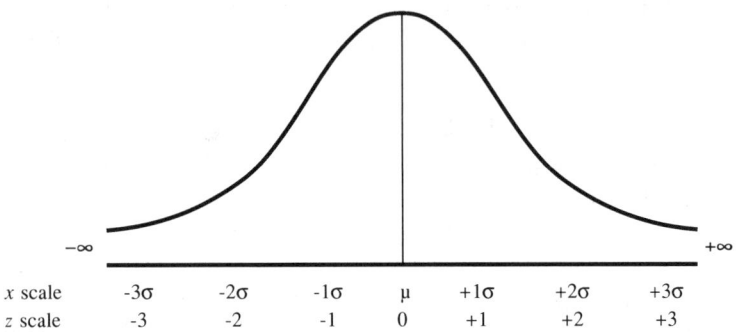

Figure 1. The Unit Normal Curve Distribution.

both above and below the mean because the normal distribution is symmetrical.

For any value, the normal curve table permits us to determine the percentage of cases falling above or below the value. For any two values, the normal curve table permits us to determine the proportion of total values or observations that fall between the two scores.

Example 1

Determining the proportion between values. If we had collected data on monthly alcohol consumption rates among college students and then converted the data to z scores, we could find the proportion of the students falling between two of the z scores we calculated, say between 1.12, which we will call observation a, and 2.5, which we will call observation b.

To determine the area between $z_a(z_a = 1.12)$ and $z_b(z_b = 2.5)$ we look up the area in the normal curve area table (Table 1). We have to find this area indirectly, however, since the table gives areas between the mean (zero) and various z scores. The area from 0 to $z_b(2.5)$ is .4938. (Look up the value for $z = 2.5$ by finding in the left column $z = 2.5$ and following that row across to the proper decimal place, in this case .00 since 2.5 + .00 = 2.5. The value in this column is the z value for 2.50 or 2.5.) This means that 49.38% of the cases or values fall between the mean (0) and 2.5z scores above the mean (Figure 2).

But since we want to know the proportion of cases that fall between z_a and z_b, we need to subtract from 49.38% the proportion of cases that

Table 1
Areas under the Normal Curve.

%	.00	.01	.02	.03	.04	.05	.06	.07	.08	.09
0.0	.0000	.0040	.0080	.0120	.0160	.0199	.0239	.0279	.0319	.0359
0.1	.0398	.0438	.0478	.0517	.0557	.0596	.0636	.0675	.0714	.0753
0.2	.0793	.0832	.0871	.0910	.0948	.0987	.1026	.1064	.1103	.1141
0.3	.1179	.1217	.1255	.1293	.1331	.1368	.1406	.1443	.1480	.1517
0.4	.1554	.1591	.1628	.1664	.1700	.1736	.1772	.1808	.1844	.1879
0.5	.1915	.1950	.1985	.2019	.2054	.2088	.2123	.2157	.2190	.2224
0.6	.2257	.2291	.2324	.2357	.2389	.2422	.2454	.2486	.2517	.2549
0.7	.2580	.2611	.2642	.2673	.2704	.2734	.2764	.2794	.2823	.2852
0.8	.2881	.2910	.2939	.2967	.2995	.3023	.3051	.3078	.3106	.3133
0.9	.3159	.3186	.3212	.3238	.3264	.3289	.3315	.3340	.3365	.3389
1.0	.3413	.3438	.3461	.3485	.3508	.3531	.3554	.3577	.3599	.3621
1.1	.3643	.3665	.3686	.3708	.3729	.3749	.3770	.3790	.3810	.3830
1.2	.3849	.3869	.3888	.3907	.3925	.3944	.3962	.3980	.3997	.4015
1.3	.4032	.4049	.4066	.4082	.4099	.4115	.4131	.4147	.4162	.4177
1.4	.4192	.4207	.4222	.4236	.4251	.4265	.4279	.4292	.4306	.4319
1.5	.4332	.4345	.4357	.4370	.4382	.4394	.4406	.4418	.4429	.4441
1.6	.4452	.4463	.4474	.4484	.4495	.4505	.4515	.4525	.4535	.4545
1.7	.4554	.4564	.4573	.4582	.4591	.4599	.4608	.4616	.4625	.4633
1.8	.4641	.4649	.4656	.4664	.4671	.4678	.4686	.4693	.4699	.4706
1.9	.4713	.4719	.4726	.4732	.4738	.4744	.4750	.4756	.4761	.4767
2.0	.4772	.4778	.4783	.4788	.4793	.4798	.4803	.4808	.4812	.4817
2.1	.4821	.4826	.4830	.4834	.4838	.4842	.4846	.4850	.4854	.4857
2.2	.4861	.4864	.4868	.4871	.4875	.4878	.4881	.4884	.4887	.4890
2.3	.4893	.4896	.4898	.4901	.4904	.4906	.4909	.4911	.4913	.4916
2.4	.4918	.4920	.4922	.4925	.4927	.4929	.4931	.4932	.4934	.4936
2.5	.4938	.4940	.4941	.4943	.4945	.4946	.4948	.4949	.4951	.4952
2.6	.4953	.4955	.4956	.4957	.4959	.4960	.4961	.4962	.4963	.4964
2.7	.4965	.4966	.4967	.4968	.4969	.4970	.4971	.4972	.4973	.4974
2.8	.4974	.4975	.4976	.4977	.4977	.4978	.4979	.4979	.4980	.4981
2.9	.4981	.4982	.4982	.4983	.4984	.4984	.4985	.4985	.4986	.4986
3.0	.4987	.4987	.4987	.4988	.4988	.4989	.4989	.4989	.4990	.4990
3.1	.49903									
3.2	.49931									
3.3	.49952									
3.4	.49966									
3.5	.49977									
3.6	.49984									
3.7	.49989									
3.8	.49993									
3.9	.49995									
4.0	.50000									

Source: James L. Bruning and B. L. Kintz, *Computational Handbook of Statistics*, Glenview, Illinois: Scott, Foresman and Company, 1968, p. 217.

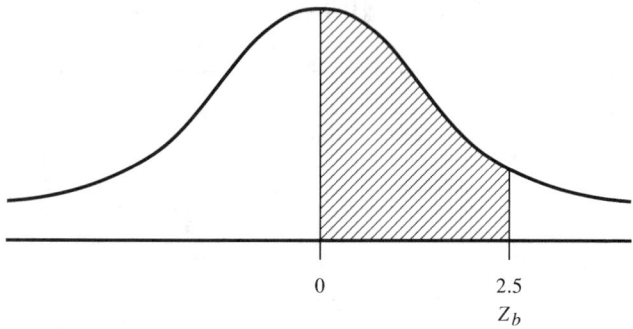

Figure 2. The Area from 0.0 to z_b.

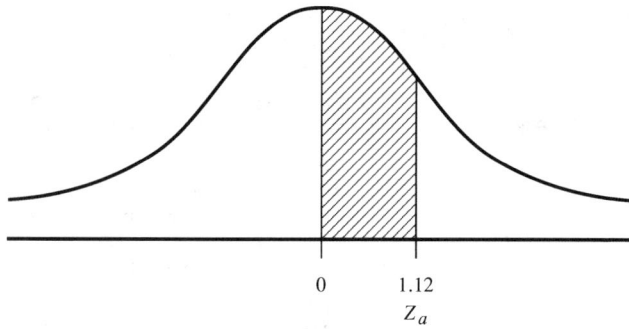

Figure 3. The Area from 0.0 to z_a.

fall between the mean and z_a. We do this by looking up the area from 0 to z_a and subtracting this from the area from 0 to z_b. (For the value of z_a = 1.12, find z = 1.1 in the left column, follow that row across to the column .02 since 1.1 + .02 = 1.12. The value at the intersection of the row 1.1 and the column .02 (.3686) is the value for z = 1.12.) The area from 0 to z_a is shown graphically in Figure 3.

The process to find the area between z_a and z_b is summarized below:

- Area 0 to z_b = .4938,
- Area 0 to z_a = .3686.

And, by subtraction:

- Area z_a to z_b = .1252 (see Figure 4).

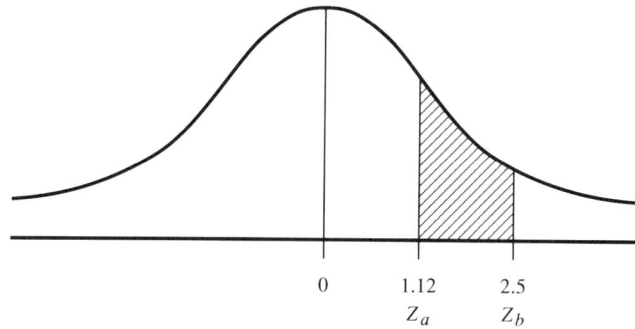

Figure 4. The Area between z_a and z_b.

(Note: The areas are obtained from Table 1, Areas under the Normal Curve. The table gives absolute values that are for areas both above and below the mean. Note also that the style and layout of these tables varies from text to text.) ∎

Example 2

Determining the proportion above and below values. The normal curve table can be useful in making estimates from data that we believe are approximately normally distributed. If we have data on assessed values for housing units for a part of the city, we could use the normal curve table to estimate the proportion of households living in units with assessed values above or below a given amount. For example, assume we have sample data for 50 of the blocks in our town, and we calculate the mean assessed value to be $14,312 and the standard deviation for the sample to be $2,200. We want to estimate the proportion of households in our town that live in dwelling units valued at $10,000 or less. We would do it as follows:

- sample of 50 blocks,
- mean assessed value of housing unit = $14,312,
- σ = $2,200 (estimated from our sample).

What proportion of households lives in dwelling units assessed at $10,000 or less? We apply the z score formula, where x_i is the particular value we are interested in ($10,000), μ is the mean assessed value

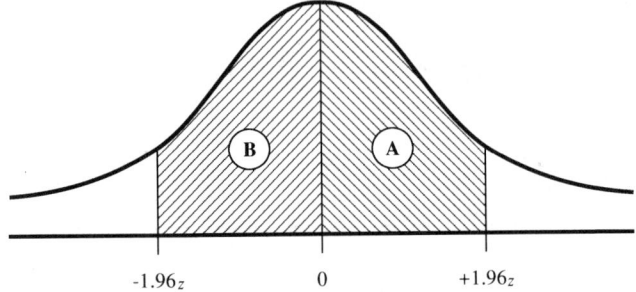

Figure 5. The Symmetry of the Normal Curve.

of $14,312 estimated from the sample, and σ is the population standard deviation of $2,200 estimated from the sample:

$$z_i = \frac{x_i - \mu}{\sigma}$$

$$= \frac{10{,}000 - 14{,}312}{2{,}200} = -1.96.$$

Since the normal curve is symmetrical, the area from 0.0 to −1.96 (**B**) is the same as the area from 0.0 to +1.96 (**A**). This is depicted in Figure 5.

Because in this case we are interested in the proportion of cases falling below a given value, we need to find the proportion of cases falling above this value and subtract the answer from 1.00. The desired area is shown in Figure 6 as area **C**.

The steps to find the desired area are as follows:

- area under upper one-half of the curve (area **D**) = .5000,
- area 0 to 10,000 (the z score of −1.96 = area **B**) = .4750,
- area *above* z_i(area **B** + **D**) = .9750, and
- area *below* z_i(area **C**) = .0250 (1.0 - .9750).

(Note: Alternatively we can subtract .4750 from .5000 because we are seeking the area in the tail of one-half of the distribution.) Thus, we estimate that 2.5% of households live in dwelling units assessed at $10,000 or less.

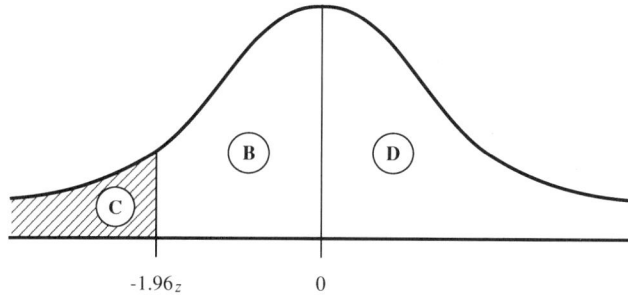

Figure 6. The Areas above and below $z = -1.96$.

This example was intended to show that the normal curve can be used to estimate the proportion of scores above, below, or between certain values. In Chapter 13 we will relate this property of the normal curve to the question of sample size and confidence level, but first we need to understand the relationship of a sample to the population. ∎

How a Sample Is Related to the Population

We select a sample and analyze its characteristics in order to derive information about the population. We want the sample from a population to reflect as closely as possible how the variable occurs in the population. When the population contains little variation, samples drawn from it will also show little variation. Consequently, the sample mean and population mean are likely to be quite similar. When the population contains great variation, so will the sample, and thus the sample mean will be less likely to be close to the population mean. The larger the sample, however, the less likely there will be a difference between the sample and population means (Caulcott, 1973, p. 34–35).

From a particular population, we could select many samples of various sizes. Grouping together all the same size samples, we could construct sampling distributions. The distributions of the means from the larger samples would have smaller standard deviations. As sample size increases, variation decreases.

If an infinitely large number of random samples of the same size were drawn from a population and their means were computed, the resulting sampling distribution of the means would closely approximate a normal distribution (Babbie, 1973, p. 85). The normal distribution would have a mean equal to the population mean (μ). Its standard deviation will equal ($\sigma/\sqrt{n}$), where σ is the population standard deviation and n is the sample size. The more random samples taken to produce the sampling distribution, the closer the sample mean will be to the population mean.

A normal distribution is determined by its mean and standard deviation (Caulcott, 1973, p. 42). Thus, the properties of the sampling distribution can be determined from its population characteristics, and we can compute the z score for the sampling distribution as follows:

$$z_i = \frac{x_i - \mu}{\sigma/\sqrt{n}}.$$

The z is therefore the difference between the original sample mean value (x_i) and the population mean (μ) in standard deviation units ($\sigma/\sqrt{n}$).

We examine the relationship between sample means and the population mean by using the properties of the normal curve distribution. As we learned earlier, the area under the normal curve represents the frequency with which a value occurs. The area under a portion of the curve tells us the probability that any element in the frequency will have a value lying between two z scores.

The properties of the normal curve are illustrated in Figure 7, which shows the proportion of the population located within various standard deviations both above and below the mean. These percentages are combined to show the proportion of cases falling within various commonly used distances above and below the mean. For example, the figure shows that 68.26% of all cases are located within one standard deviation (1SD) above and 1SD below the mean, or that only .11% of the cases are located above 3SD above the mean.

We can determine a variety of relationships from this curve, including the following that are most commonly used in practice, and which will be used later when we discuss sample size in Chapter 13:

- 68% of the sample values will lie between the population mean and ±1SD.
- 95% of the sample values will lie between the population mean and ± 2SD (actually the value is 1.96 standard deviations, but it is often rounded to 2 standard deviations for convenience).

Confidence Levels

- 99% of the sample values will lie between the population mean and ±3SD (actually the value is 2.58 standard deviations, but it is often rounded to 3 standard deviations for convenience).

Usually we do not have data on the entire population; we have only the information from a single sample. We need to know how representative that sample is of the population from which it is taken. When we do not know the population mean, we have to use the sample mean as the best estimate; if the sample has been carefully selected, the sample mean will be close to the population mean, but not necessarily the

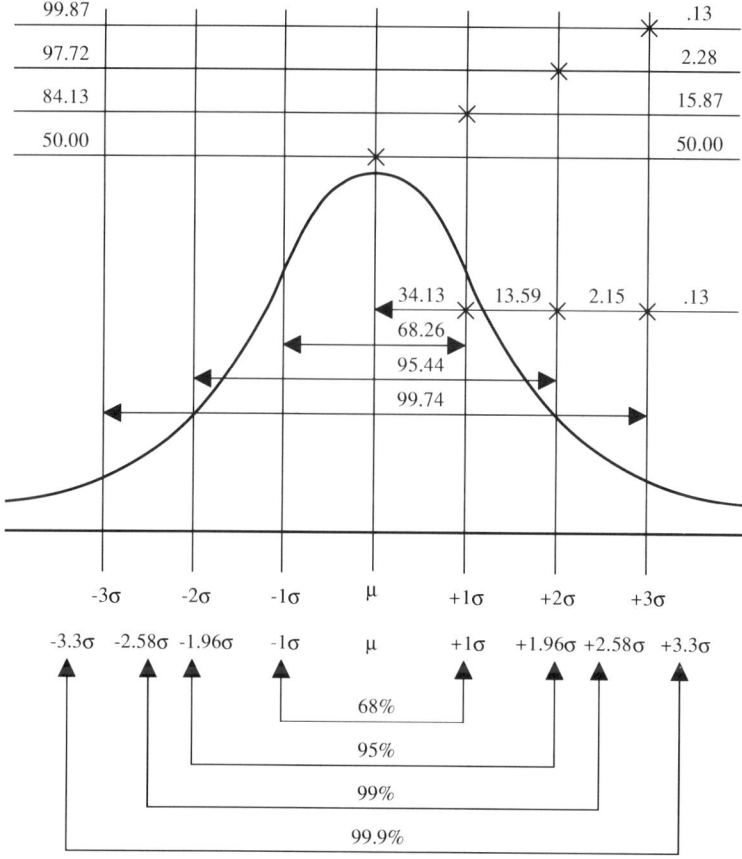

Figure 7. The Normal Curve.

same. We also know that the sample standard deviation (s) will underestimate the population standard deviation (σ) (Runyon and Haber, 1980, pp. 214–217). We use our sample standard deviation to estimate the standard deviation of the sampling distribution (σ/n). We use the value for s for σ, and by using the properties of the normal distribution we can state that based on the mean and standard deviation of the sample, the *possible population means* will fall as follows:

- 68% will lie between the sample mean and $\pm s/\sqrt{n}$.
- 95% will lie between the sample mean and $\pm 1.96\ s/\sqrt{n}$.
- 99% will lie between the sample mean and $\pm 2.58\ s/\sqrt{n}$.

Note that here and in the remaining examples we use the precise values rather than the rounded values for the number of standard deviations.

Example 3

Estimating population means. Using the example of 50 city blocks with the mean assessed value = $14,312, and the sample's standard deviation of $2,200 as an estimate of the population's standard deviation:

$$\sigma = 2,200,$$
$$s = 2,200,$$
$$s/\sqrt{n} = 2,2000/\sqrt{50} = 2,200/7.1 = 309.$$

Thus:

- 95% of the possible population means fall within the range of the sample mean $\pm 1.96\ s/\sqrt{n}$,

or, of all the populations from which the sample could be drawn,

- 95% have means of $14,312 ± 1.96 ($309),

or,

- 95% have means of $14,312 ± $606 ($605.64 rounded to the nearest dollar),

or,

- 95% have means ranging from $13,706 to $14,918. ∎

Standard Error

The standard deviation of the sampling distribution measures the probable difference between the sample mean and population mean. Referring to the previous example, it tells the probable *error* in using the sample mean to estimate the population mean (Caulcott, 1973, p. 49). Thus, the standard deviation of the sampling distribution is called the *standard error* or the *standard error of the mean*. It is a measure of the error of the sample *statistic*, in this case the mean, but it can measure the error in other statistical measures as well. The standard error thus measures the error in using the sample statistic as an estimate of the population value.

In the above example, the 95% was the *confidence level*, and the ±$606 (±$605.64 rounded to the nearest dollar) was the *confidence interval*. Thus, the range of possible means (±$606) is related to the probability of being correct (95%) in inferring that the interval contains the population mean.

Example 4

Estimating representativeness. To measure how *representative a sample is of the population*, we use the standard error of the mean (SEM) as previously defined.

$$SEM = s/\sqrt{n}.$$

Assume
- a sample of 30 city blocks,
- mean number of households = 15,
- $s = 3.5$ (estimated from previous work on the topic or from the sample).

Therefore,

$$SEM = 3.5/\sqrt{30} = 3.5/5.48 = .64.$$

The SEM in this case tells us that if we were to continue taking samples:
- 68% of their means will equal 15 ± .64,
- 95% of their means will equal 15 ± 1.25,
- 99% of their means will equal 15 ± 1.65. ■

Example 5

Increasing sample size. Increasing sample size gives more precision to our estimate because the confidence intervals are narrowed. This can be seen when we compute the same information as above but assume a sample of 100 blocks rather than 30.

$$SEM = 3.5/\sqrt{100} = 3.5/10 = .35.$$

- 68% of their means will equal $15 \pm .35$,
- 95% of their means will equal $15 \pm .69$,
- 99% of their means will equal $15 \pm .90$. ∎

Use of Confidence Limits

The information presented earlier is needed in the following chapter when we determine the size of samples needed in order to be able to be confident about the information we obtain through surveys, and in order to check the work of others. The concepts presented earlier are useful in themselves, however, as they permit us to make estimates about the proportions of populations that exhibit certain characteristics. The usefulness of confidence levels and confidence limits will be more evident after the next chapter has been mastered.

p-Value

In Chapter 6, Statistical Significance, we mentioned the p-value test as an alternative approach to setting a significance level for inferring whether a difference was due to chance. Now that we know about the normal curve, the standard error, confidence levels, etc., we can use the p-value to test the null hypothesis.

Example 6

Applying the p-value. Assume that we have experienced very rapid growth in our town during the past year and believe that the average household income of new residents is substantially greater than that of the average household income of pre-existing residents. Last year's average household income, when adjusted for inflation to current dollars, is $24,000 with a standard deviation of $3,000. We took a sample of 100 new households and found their average household income to be $24,500. This difference in average household incomes looks important, but is this difference only due to sampling error? Or, is the average household income of the new residents statistically different than last year's average? (For this example, assume that the number of households moving out of town during the past year was insignificant.)

The null hypothesis (H_o) is that there is no difference between the average household incomes. The alternative hypothesis (H_1) is that there is a difference. The p-test allows us to determine how consistent the sample ($24,500) is with the null hypothesis ($24,000) (Wonnacott and Wonnacott, 1984, pp. 262–263). If the null hypothesis were true, what is the probability that the sample value would be a high as $24,500?

We calculate the z score for the difference between the sample value and the null hypothesis.

$$z_i = \frac{x_i - \mu}{\sigma\sqrt{n}},$$

$$= \frac{24,500 - 24,000}{3,000/\sqrt{100}},$$

$$= \frac{500}{300} = 1.667.$$

We look up the z score in the normal table as we have done before to determine the probability of finding a sample value as high as the one we did. We are looking for the percentage of the distribution that is in the upper tail. Because our normal curve table does not present this information directly, we need to find the area in the tail indirectly as we did earlier:

- area under one-half of the curve = .5000,
- area from $24,000 to $24,500 (the z score of 1.67) = .4525,
- area below $24,500 = .9525,
- area above $24,500 = .0475 (1.0 - .9525).

(Note: Alternatively .4525 can be subtracted from .5000 because we are looking for the area in the tail of one-half of the distribution.)

If the new households had incomes no higher than the old residents (if H_o were true), then there would be a 4.75% probability of observing a sample household average income as large as $24,500 (Figure 8). This p-value of 4.75% then tells us how much agreement there is

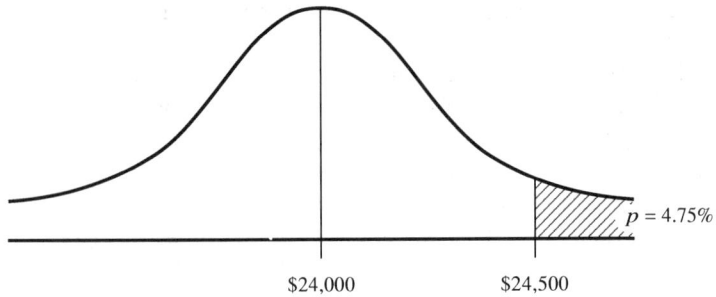

Figure 8. The *p*-Value of 4.75%.

between the data and the null hypothesis. Here the data provide very little statistical support for the null hypothesis. It is up to the reader or analyst to make such a conclusion, but usually we would say that the 4.75% does not provide enough support for the null hypothesis, so we reject it. We would, therefore, interpret these data by saying that new residents very likely have statistically significantly higher incomes than old residents. As the analyst, you will need to determine whether this $500 statistically significant difference is an *important* or *meaningful difference*.

Note that if the sample household income of new residents had been higher, there would have been even less of a probability that the data support the null hypothesis. ■

Appendix: How to Calculate the Standard Deviation for a Sample

Assume that we have a large school with many classrooms of students and we want to compute the standard deviation of students per classroom. We select a sample of classrooms, lay out the data, shown in the table following, and compute the sample mean, which turns out to be 15. Then for each classroom we find the deviation from the mean by comparing the number of students in the room with the mean and finding the positive or negative result. We then square that result to get the squared deviation and insert these results into the formula for the standard deviation of a sample.

Confidence Levels

1. Lay out the basic data:

Classroom	Number Students	Deviation from Mean	Squared Deviation
4	20	5	25
3	15	0	0
2	10	–5	25
1	15	0	0
	60	10	50

Mean = 60/4 = 15.

2. Insert the data into the formula:

SD = $\sqrt{\text{sum of the squared deviations/total number of cases} - 1}$

or, in mathematical symbols,

$$s = \sqrt{\sum (x^2)/n-1},$$

where

s = the standard deviation (SD) for a sample,
x = deviation from the arithmetic mean, and
n = total number of items (in this case, classrooms).

$$s = \sqrt{(50/3)} = \sqrt{16.67} = 4.08.$$

Note that this formula is for the standard deviation for a sample. We use this as an estimate of the standard deviation for the population when we are not able to obtain the standard deviation for the population. The formula for the standard deviation for the population is $\sigma = \sqrt{\Sigma(x^2)/N}$ (see Chapter 1, Descriptive Statistics). Because the computation is for the population, we do not subtract 1 as in the computation for the standard deviation for the sample.

Further Readings

Babbie, Earl R., *Survey Research Methods*. Belmont, California: Wadsworth, 1973.

Bruning, James L., and B. L. Kintz, *Computational Handbook of Statistics*. Glenview, Illinois: Scott, Foresman and Company, 1968.

Caulcott, Evelyn, *Significance Tests*. London: Routledge and Kegan Paul, 1973.
Matlack, William F., *Statistics for Public Policy and Management*. Boston: Duxbury Press, 1980.
Runyon, Richard P., and Audrey Haber, *Fundamentals of Behavioral Statistics, Fourth Edition*. Reading, Massachusetts: Addison-Wesley, 1980.
Wonnacott, Ronald J., and Thomas H. Wonnacott, *Statistics: Discovering Its Power*. New York: John Wiley and Sons, 1982.
Wonnacott, Thomas H., and Ronald J. Wonnacott, *Introductory Statistics for Business and Economics, Third Edition*. New York: John Wiley and Sons, 1984.

Chapter 13
SAMPLE SIZE

Definition

The size of sample taken affects the extent to which we can place confidence in the data collected through a survey. All other things being equal (the samples were selected properly, the data were collected consistently, they were recorded accurately, etc.), a larger sample will yield statistics that are more representative of the actual values in the population than will a smaller sample. But since costs increase with sample size, we want to take a sample that is only as large as necessary to produce the level of accuracy needed. The two key questions to be asked are (1) how accurate do the data need to be, and (2) how much time and money do we have available?

Method

In practice, there are two principal methods used to determine sample size. We can imitate the size of samples taken by others that yielded what are generally considered to be good results, or we can base our sample size on a determination of the level of accuracy we believe is necessary for the concept or problem we are investigating. We will focus on this second method, but we have also appended summary tables that allow us to estimate the size of sample needed.

Using the information in Chapter 12 about the normal curve, the standard error, confidence intervals, and confidence levels, we can determine how large a sample should be taken in order to assure a given level of accuracy.

We will describe four methods for computing sample size, the first two using the previously given information.

1. Estimating the sample size for a mean.
2. Estimating sample size for a percentage.
3. Estimating sample size to produce a minimum cell size for a cross tabulation.
4. Estimating the size sample needed to produce a specific difference from chance in a cross tabulation.

Example 1a

Estimating the sample size for a mean. We wish to estimate the number of rooms per unit (or number of persons per unit) as part of a density study related to changing family composition. We ask the two key questions:

- What confidence level do we need?
- What confidence interval do we need?

At issue is the number of rooms per unit.

We answer, for example, that we wish to have a sample large enough to achieve the 95% confidence level and have our estimate be within one-half of a room. So, we set the following parameters:

- 95% confidence level. (See Chapter 12, Confidence Levels, if you do not understand this concept.)
- ±.5 rooms (a confidence interval of 1.0) of the true mean. (See Chapter 12 if you do not understand this concept. Note that a confidence interval of ±.5% is the same as a confidence interval of 1.0.)

We calculate the sample size by multiplying the standard error of the mean (SEM) by the z score corresponding to the selected confidence

Sample Size

interval and equating this to one-half of the length of the required confidence interval. (See the information presented in Chapter 12.)

$$CI/2 = Z_* \, \sigma/\sqrt{n},$$

where

CI = the Confidence Interval,
Z_* = the z score corresponding the selected confidence level,
CI/2 = one–half of the confidence interval corresponding to the required confidence level, and
$\sigma/\sqrt{n}$ = the standard error of the mean.

Assume $\sigma = 1.5$ (from previous work on the topic). (See Chapter 1, Descriptive Statistics, for advice on calculating σ, the standard deviation.)

$$CI/2 = (1)(\sigma/\sqrt{n}) \quad \text{[at 68\% level]},$$

$$CI/2 = (2)(\sigma/\sqrt{n}) \quad \text{[at 95\% level, so we multiply by 2]},$$

$$.5 = 2(\sigma/\sqrt{n}) \quad \text{[.5 is one–half of the confidence interval — ±.5 or 1.0 room]},$$

$$.5 = 2(1.5/\sqrt{n}).$$

[Note that the goal here is to use algebra to isolate n on one side of the equation. If we square the above equation, this process removes the radical sign and we get the following equation.]

$$.25 = 4 \, (2.25/n),$$
$$.25 = 9/n,$$
$$.25n = 9/n,$$
$$n = 9/.25,$$
$$n = 36.$$

We need a sample of 36 units.
Check:

$$CI/2 = (2) \, 1.5/\sqrt{36}$$
$$= (2) \, 1.5/6$$
$$= (2) \, (.25)$$
$$= .5.$$

The above formula can be rewritten so that we can solve directly for n.

$$n = \left[\frac{Z_* \sigma}{CI/2}\right]^2$$

$$= \left[\frac{2 \times 1.5}{.5}\right]^2$$

$$= \left[\frac{3}{.5}\right]^2$$

$$= 6^2$$

$$= 36.$$

Note that in this and the following examples we are speaking about *achieved* sample sizes. When you expect less than a 100% response, you will need to take a larger sample in order to compensate for the less than 100% response. ■

Example 1b

Estimating the sample size for a mean when there is a large standard deviation. When the variation in a set of data is great, we will need a larger sample to assure that the values we determine by sampling reflect accurately the values in the population. In this case, assume we wish to determine the average number of households on a city block and we know that the number of households per city block ranges from only a few to very many. At issue here is the average number of households per block. We then ask:

- What confidence level do we need?
- What confidence interval do we need?

Answer, for example:

- 95% confidence level (only 5% chance of error).
- We want to be within ± .3 households (confidence interval of .6) of the true mean; that is, we want a close estimate.

$$CI/2 = Z_* \sigma/\sqrt{n} \; ;$$

Sample Size

at 95%,
$$CI/2 = (2)(\sigma/\sqrt{n}).$$

Assume $\sigma = 3.5$ (from previous work on the topic):

$$.3 = 2(\sigma/\sqrt{n}) = 2(3.5/\sqrt{n}),$$
$$.09 = 4(12.25/n),$$
$$.09 = 49/n,$$
$$.09n = 49,$$
$$n = 49/.09,$$
$$n = 544.44$$

We should take a sample of 545 blocks (rounding up).
Check:

$$CI/2 = (2)\frac{3.5}{\sqrt{544.44}}$$
$$= (2)\frac{3.5}{23.33}$$
$$= .3.$$

Solving directly for n:

$$n = \left[\frac{Z_*\sigma}{CI/2}\right]^2$$
$$= \left[\frac{2 \times 3.5}{.3}\right]^2$$
$$= \left[\frac{7}{.3}\right]^2$$
$$= 23.3333^2$$
$$= 544.44$$

■

Example 2a

Estimating sample size for a percentage. In this case we wish to know the percentage of voters who are in favor of or opposed to a proposal.
The formula for the standard error of a percentage is

$$SE = \sqrt{(PQ/n)},$$

where

P = percentage in favor,
$Q = 1-P$, or the percentage opposed, and
n = number of cases in the sample.

Assume that in a survey of voter preference we found a 60/40 split in a sample of 200 persons (i.e., 60% in favor, 40% opposed):

$$SE = \sqrt{(.60 \times .40/200)} = \sqrt{(.24/200)} = \sqrt{.0012} = .0346 \text{ or } 3.5\%.$$

We would be 68% confident that the sample estimate is within one standard error of the population parameter, that is, 60% ± 3.5% and 40% ± 3.5%.

We would be 95% confident that the sample estimate was within two standard errors, that is 60% ± 7% and 40% ± 7%, or between 53% to 67% and 33% to 47%.

Calculating the sample size for a percentage:

- Ask—confidence level?
- Ask—confidence interval?

In a straw poll on an issue last year we found 80% pro, 20% con. We want to work at the 95% confidence level and are willing to be off 5 percentage points either way (the confidence interval).

$$CI/2 = Z_* \sqrt{(PQ/n)},$$

where

Z_* = the z score corresponding the selected confidence level,
$CI/2$ = one-half of the confidence interval corresponding to the required confidence level, and
$\sqrt{PQ/n}$ = the standard error of the percentage.

Sample Size

$$CI/2 = 2\sqrt{(PQ/n)},$$
$$.05 = 2\sqrt{(.80 \times .20/n)},$$
$$.05 = 2\sqrt{(.16/n)},$$
$$.0025 = 4(.16/n),$$
$$.0025n = 64,$$
$$n = .64/.0025,$$
$$n = 256.$$

Thus, we would need a sample of 256 persons in order to obtain results that would be within ± 5 percentage points at the 95% confidence level.

This formula can also be rewritten so that we solve directly for n. Using the notation system from above, the formula would be

$$n = \left[\frac{Z_*\sqrt{PQ}}{CI/2}\right]^2.$$

The formula can be rewritten in the following form, which allows for easier algebraic solutions:

$$n = \frac{Z_*^2 PQ}{(CI/2)^2}.$$

Both solutions are shown for this example, but we will use the second formula for later problems.

First formula:

$$n = \left[\frac{Z_*\sqrt{PQ)}}{CI/2}\right]^2$$
$$= \left[\frac{2\sqrt{(.80 \times .20)}}{.05}\right]^2$$
$$= \left[\frac{2\sqrt{(.16)}}{.05}\right]^2$$
$$= \left[\frac{2 \times .4}{.05}\right]^2$$
$$= 16^2$$
$$= 256.$$

Second formula:

$$n = \frac{Z_*^2 PQ}{(CI/2)^2}$$

$$= \frac{2^2(.80 \times .20)}{.05^2}$$

$$= \frac{4(.16)}{.0025}$$

$$= .64/.0025$$

$$= 256.$$

■

Example 2b

Estimating sample size for a percentage, but at a higher confidence level. If we wanted to take a sample large enough to be within ±5 percentage points at the 99% level, we would compute it as follows:

$$CI/2 = 3\sqrt{(PQ/n)},$$

$$.05 = 3\sqrt{(.80 \times .20/n)},$$

$$.05 = 3\sqrt{(.16/n)},$$

$$.0025 = 9(.16/n),$$

$$.0025n = 1.44,$$

$$n = 1.44/.0025,$$

$$n = 576;$$

or, solving for n:

$$n = \frac{Z_*^2 PQ}{(CI/2)^2}$$

$$= \frac{3^2(.16)}{.05^2}$$

$$= \frac{9(.16)}{.0025}$$

Sample Size

$$= \frac{1.44}{.0025}$$

$$= 576.$$

If you have no idea of the percentages, you can make a conservative estimate, since we know that the product PQ reaches the maximum where $P = Q = .50$.

Recalculating Example 2a with a 50/50 split, ±5% at the 95% level:

$$.05 = 2\sqrt{(.50 \times .50/n)},$$

$$.05 = 2\sqrt{(.25/n)},$$

$$.0025 = 4(.25/n),$$

$$.0025n = 1.0,$$

$$n = 1.0/.0025,$$

$$n = 400;$$

or, solving for n:

$$n = \frac{Z_*^2 PQ}{(CI/2)^2}$$

$$= \frac{2^2(.25)}{.05^2}$$

$$= \frac{4(.25)}{.0025}$$

$$= \frac{1.0}{.0025}$$

$$= 400.$$

For convenience, we have summarized below in Table 1 the sample sizes that would result when we expect a 50/50 split in responses related to various confidence levels and confidence intervals. See also Table 5 in the appendix to this chapter. It can be used to estimate the sample size needed for responses split at various percentages between 50/50 and 90/10 and at various confidence intervals at the 95% confidence level. Note that these are *achieved* sample sizes. In order to achieve a sample of 100 through a mailed survey, for example, we need to distribute more than 100 questionnaires to compensate for a less than 100% response rate. If we expected an 80% response rate we would have to distribute 125 questionnaires to receive 100 returns. ∎

Table 1

Sample Sizes for Selected Confidence Levels and Intervals
When We Expect a 50/50 Split in Responses

Confidence Level	Confidence Interval	Achieved Sample Size
68%	±5.0%	100
68%	±2.5%	400
95%	±10.0%	100
95%	±5.0%	400
95%	±4.0%	625
95%	±2.5%	1600
99%	±10.0%	225
99%	±5.0%	900
99%	±2.5%	3600

Example 3

Estimating the size of sample needed to produce a minimum cell size for a tabular analysis. Sometimes we want to be sure we select a sample that will be large enough to enable us to do a sub-cell/sub-category analysis. For example, we want to take a sample of surgeons that will contain at least a certain number of female brain surgeons (or a sample of poor people who own the homes in which they live). If we are sampling from a population that is not stratified, we will have to take a random sample of the entire population that is large enough to yield the minimum number for the sub-category in question. We can do this by using the minimum cell size formula when we are able to estimate the marginal splits of the subject variables. We use the following formula:

$$n \text{ [minimum sample size]} = \frac{\text{minimum cell size}}{(\text{marginal \%})(\text{marginal \%})}.$$

Assume in the example above that we want to maximize our chance of having at least five female brain surgeons in our sample. Assume

Sample Size

also that 20% of surgeons are females and 30% of surgeons are brain surgeons. We would estimate the sample size as shown below:

$$n = \frac{5}{(.3)(.2)}$$

$$= \frac{5}{.06}$$

$$= 83.33$$

Thus, in a sample of 84 surgeons (83.33 rounded up), the most likely number of female brain surgeons would be 5.

Note that since we want the smallest sample size, we work with the two smallest marginal percentages. In the above example they were 70/30 and 80/20 so we use 30% and 20%. Table 2 displays the estimated results.

Table 2

Distribution of Surgeons

	Surgeons		
	Brain	Other	Total
Male	20	47	67 (80%)
Female	5	12	17 (20%)
	25 (30%)	59 (70%)	84

If we have variables with more than two values, the formula is applied to the smallest marginal percentage for each variable. For example, if one variable were split 50/40/10 and the other were split 65/20/15, we would use the two smallest marginals, 10% and 15%. Assume we wish to have a minimum cell size of 10; we would determine the minimum sample size as:

$$n = \frac{10}{(.10)(.15)}$$

$$= \frac{10}{.015}$$

$$= 667.$$

Table 3
Minimum Achieved Sample Sizes for a Minimum Cell Size of 5

Marginal Split on y	Marginal Split on x				
	50/50	70/30	90/10	95/5	99/1
50/50	20	34	100	200	1,000
70/30	34	56	167	334	1,667
90/10	100	167	500	1,000	5,000
95/5	200	334	1,000	2,000	10,000
99/1	1,000	1,667	5,000	10,000	50,000

Note: We would use a different, more efficient sampling procedure when we have very small marginal percentages, e.g., less than 10% on either x or y (see Sudman, 1976, pp. 25–47).

For convenience, the minimum sample sizes for various marginal percentages are summarized in Table 3 for a minimum cell size of 5.

■

Example 4a

Using χ^2 to Estimate Sample Size. (This discussion assumes you understand the concept of χ^2. If you do not, see Chapter 6, Statistical Significance.) Assume that we want to select a sample that is large enough to assure that specified differences from chance will be found statistically significant.

Follow these steps, assuming we are examining data about the preferences of new and old residents of our town toward industrial development incentives. From earlier surveys we learned that 60% of the inhabitants are long-term or older residents and 40% are newer residents. We also found that 30% of the residents oppose new industry and 70% are in favor of it.

1. Specify the level of significance (e.g., .05; this is equivalent to the 95% confidence level discussed earlier).
2. Specify the least difference you want to be detected as significant (e.g., a 20 percentage point difference found significant at the .05 level).
3. Produce expected and least difference tables. Construct the table of expected proportions from the data we have on hand as shown in Table 4a.

Table 4a

Expected Proportions: Position of Residents Toward Industry

Resident Status	Position on Industry		
	No Ind.	Ind.	
New	.12	.28	.40
Old	.18	.42	.60
	.30	.70	1.00

Source: Existing data for Our Town.

Note: The cell proportions are the product of the marginals. For example, New/No Ind. = .40 × .30 = .12 .

Table 4b

Least Difference Table (20% difference)

	No Ind.	Ind.	
New	.22	.18	.40
Old	.08	.52	.60
	.30	.70	1.00

Construct a least difference table by applying the percentage difference we expect to find to the diagonal values of Table 4b. If we are seeking a 20% difference, we add 10 percentage points to one value in the row and subtract 10 percentage points from the other value in the row. We reverse the procedure for the second row. Note that the row and column marginals must retain the same proportional relationships.

4. Enter the least-difference proportions (using them as observed frequencies) and the expected proportions in the χ^2 formula.
5. Solve for n. In this case set χ^2 equal to 3.84 (3.84 is the χ^2 value for 1 d.f. at the .05 level). (See Table 8 in Chapter 6, Statistical Significance.)

$$\chi^2 = \sum \frac{(O-E)^2}{E},$$

where

O = observed proportions,
E = expected proportions, and
Σ = summation.

$$\chi^2 = 3.84 = \frac{(.22n - .12n)^2}{.12n} + \frac{(.18n - .28n)^2}{.28n}$$

$$+ \frac{(.08n - .18n)^2}{.18n} + \frac{(.52n - .42n)^2}{.42n}$$

$$= 3.84 = \frac{(.10n)^2}{.12n} + \frac{(-.10n)^2}{.28n} + \frac{(-.10n)^2}{.18n} + \frac{(.10n)^2}{.42n}$$

$$= 3.84 = \frac{.01n^2}{.12n} + \frac{.01n^2}{.28n} + \frac{.01n^2}{.18n} + \frac{.01n^2}{.42n}$$

$$= 3.84 = .083n + .036n + .056n + .024n$$

$$= 3.84 = .199n,$$

$$n = 3.84/.199 = 19.30.$$

Thus, we need a sample of 20 persons (19.30 rounded up) to be sure of detecting a 20% difference from chance 95 out of 100 times. When determining sample size, always round up to help assure a sample that is sufficiently large.

We can check this estimating procedure by using the expected and least difference proportions to compute χ^2 to determine whether it has the value of 3.84. This is shown in Tables 4c1 through 4d2.

To produce the expected and least difference values for the χ^2 formula, we compute the cell frequencies by multiplying the total number

Table 4c1

Expected Proportions

.12	.28	
.18	.42	
		1.0

Table 4c2

Least Difference Proportions

.22	.18	
.08	.52	
		1.0

Sample Size

Table 4d1
Expected Frequencies

2.32	5.40	
3.47	8.11	
		19.3

Table 4d2
Least Difference Frequencies

4.25	3.47	
1.54	10.04	
		19.3

of cases (19.3) by each cell proportion (for example, 19.3 × .12 = 2.32). Note that there will be rounding errors when we compute the cell frequencies.

$$\chi^2 = \frac{(4.25 - 2.32)^2}{2.32} + \frac{(3.47 - 5.40)^2}{5.40}$$
$$+ \frac{(1.54 - 3.47)^2}{3.47} + \frac{(10.04 - 8.11)^2}{8.11}$$
$$= \frac{(1.93)^2}{2.32} + \frac{(-1.93)^2}{5.40} + \frac{(-1.93)^2}{3.47} + \frac{(1.93)^2}{8.11}$$
$$= 1.606 + .690 + 1.074 + .459$$
$$= 3.83 \quad \text{(which is not 3.84 because of rounding errors).}$$

■

Example 4b

Using the same data, but we want to assure a 10% difference at the .05 level. Note that when we want to detect a smaller difference we must take a larger sample. If this is not intuitively clear, the example given in Tables 4e1 and 4e2 demonstrates this principle.

Table 4e1
Relative Proportions

	No Ind.	Ind.,	
New	.12	.28	.40
Old	.18	.42	.60
	.30	.70	1.00

Table 4e2
Least Difference Proportions (10% Difference)

	No Ind.	Ind.	
New	.17	.23	.40
Old	.13	.47	.60
	.30	.70	1.00

$$\chi^2 = 3.84 = \frac{(.17n - .12n)^2}{.12n} + \frac{(.23n - .28n)^2}{.28n}$$

$$+ \frac{(.13n - .18n)^2}{.18n} + \frac{(.47n - .42n)^2}{.42n}$$

$$= 3.84 = \frac{(.05n)^2}{.12n} + \frac{(-.05n)^2}{.28n} + \frac{(-.05n)^2}{.18n} + \frac{(.05n)^2}{.42n}$$

$$= 3.84 = \frac{.0025n^2}{.12n} + \frac{.0025n^2}{.28n} + \frac{.0025n^2}{.18n} + \frac{.0025n^2}{.42n}$$

$$= 3.84 = .021n + .009n + .014n + .006n$$

$$= 3.84 = .05n,$$

$$n = 3.84/.05 = 76.80.$$

Thus, we need a sample of 77 persons to be sure of detecting a 10% difference from chance 95 out of 100 times. (Because a 10% difference is more likely to occur by chance than a 20% difference, we need a larger sample to be more confident that the 10% difference we detect did not occur by chance.)

We can also check this estimate by using the expected and least difference proportions to compute χ^2 to determine whether it has the value of 3.84. This is shown in Tables 4f1 through 4g2. To produce the expected and least difference values for the χ^2 formula, we compute the cell frequencies by multiplying the total number of cases (76.82) by each cell proportion (for example, 76.82 × .28 = 21.51). Note that there will be rounding errors when we compute the cell frequencies and our final results will not exactly equal the χ^2 value.

$$\chi^2 = \frac{(13.06 - 9.22)^2}{9.22} + \frac{(17.67 - 21.51)^2}{21.51}$$

$$+ \frac{(9.99 - 13.83)^2}{13.83} + \frac{(36.11 - 32.26)^2}{32.26}$$

$$= 1.60 + .686 + 1.07 + .460$$

$$= 3.82 \quad \text{(which is not 3.84 because of rounding errors).}$$

Sample Size 167

Table 4f1		
Expected proportions		
.12	.28	.40
.18	.42	.60
.30	.70	1.00

Table 4f2		
Least Difference Proportions		
.17	.23	.40
.13	.47	.60
.30	.70	1.00

Table 4g1		
Expected Frequencies		
9.22	21.51	
13.83	32.26	
		76.82

Table 4g2		
Least Difference Frequencies		
13.06	17.67	
9.99	36.11	
		76.82

How Large a Sample is Sufficient?

The foregoing methods can be used to estimate the size of sample needed to help assure that the information we collect through a sample is a good approximation of the true value in the population. While there always is some error associated with the statistics we garner through sampling, the methods in this chapter can help us determine how large a sample should be taken and how much confidence we can place in the results. Remember that you often have to take a larger sample than the ones identified in this chapter in order to compensate for a less than 100% response rate.

The size of the population from which we are sampling also influences the size of sample we must take. In this chapter, for simplicity's sake, we assumed that we were sampling from a large population (e.g., 100,000 persons or more). When sampling from smaller populations, the sample size can be reduced somewhat but allow us to retain the same confidence level and confidence interval. Table 6 summarizes sample sizes that may be used for smaller populations, at the 95% confidence level and for confidence intervals from ±1% to ±10%.

Table 7 summarizes common sample sizes used for various types of research at the regional and national levels. National studies tend to have samples of 1000 or more, while regional studies have smaller

Table 5

Estimated Sampling Error for a Binomial (95% Confidence Level).

	BINOMIAL PERCENTAGE DISTRIBUTION				
Sample size	50/50	60/40	70/30	80/20	90/10
100	10	9.8	9.2	8	6
200	7.1	6.9	6.5	5.7	4.2
300	5.8	5.7	5.3	4.6	3.5
400	5	4.9	4.6	4	3
500	4.5	4.4	4.1	3.6	2.7
600	4.1	4	3.7	3.3	2.4
700	3.8	3.7	3.5	3	2.3
800	3.5	3.5	3.2	2.8	2.1
900	3.3	3.3	3.1	2.7	2
1000	3.2	3.1	2.9	2.5	1.9
1100	3	3	2.8	2.4	1.8
1200	2.9	2.8	2.6	2.3	1.7
1300	2.8	2.7	2.5	2.2	1.7
1400	2.7	2.6	2.4	2.1	1.6
1500	2.6	2.5	2.4	2.1	1.5
1600	2.5	2.4	2.3	2	1.5
1700	2.4	2.4	2.2	1.9	1.5
1800	2.4	2.3	2.2	1.9	1.4
1900	2.3	2.2	2.1	1.8	1.4
2000	2.2	2.2	2	1.8	1.3

How to use this table: Find the intersection between the sample size and the approximate percentage distribution of the binomial in the sample. The number appearing at this intersection represents the estimated sampling error, at the 95% confidence level, expressed in percentage points (plus or minus).

Example: In a sample of 400 respondents, 60% answer "Yes" and 40% answer "No." The sampling error is estimated at plus or minus 4.9 percentage points. The confidence interval, then, is between 55.1% and 64.9%. We would estimate (95% confidence) that the proportion of the total population who would say "Yes" is somewhere within that interval.

Source: Earl L. Babbie, *Survey Research Methods*. Belmont, California: Wadsworth, 1973, p. 376

samples. When there are subgroup analyses, the sample sizes are increased, to assure for adequate numbers of respondents in the subgoups, as we discussed in Chapter 11, Sampling.

Tables 5, 6, and 7 are not intended to replace the formal estimation procedures presented in this chapter. Rather, they are intended to be used in initial planning and as a check on the accuracy of your estimates.

Nothing can guarantee that our results will be accurate, but these estimating methods can help reduce the risk of making gross errors. In

Table 6

Sample Size for Specified Confidence Limits and Precision When Sampling Attributes in Percent.

A. 95% confidence interval ($p = 0.5$)[a]

Population size	Sample size for precision of					
	±1%	±2%	±3%	±4%	±5%	±10%
500	b	b	b	b	222	83
1,000	b	b	b	385	286	91
1,500	b	b	638	441	316	94
2,000	b	b	714	476	333	95
2,500	b	1,250	769	500	345	96
3,000	b	1,364	811	517	353	97
3,500	b	1,458	843	530	359	97
4,000	b	1,538	870	541	364	98
4,500	b	1,607	891	549	367	98
5,000	b	1,667	909	556	370	98
6,000	b	1,765	938	566	375	98
7,000	b	1,842	959	574	378	99
8,000	b	1,905	976	580	381	99
9,000	b	1,957	989	584	383	99
10,000	5,000	2,000	1,000	588	385	99
15,000	6,000	2,143	1,034	600	390	99
20,000	6,667	2,222	1,053	606	392	100
25,000	7,143	2,273	1,064	610	394	100
50,000	8,333	2,381	1,087	617	397	100
100,000	9,091	2,439	1,099	621	398	100
$-\infty$	10,000	2,500	1,111	625	400	100

[a] p—Proportion of units in sample possessing characteristic being measured; for other values of p, the required sample size will be smaller.

[b] In these cases 50% of the universe in the sample will give more than the required accuracy. Since the normal distribution is a poor approximation of the hypergeometrical distribution when x is more than 50% of N, the formula used in this calculation does not apply.

Source: Taro Yamane, *Elementary Sampling Theory*, Englewood Cliffs, New Jersey: Prentice-Hall, Inc., 1976, p. 398.

practice, samples are often smaller than desired because of time and money constraints, and as a consequence, their results must be interpreted cautiously. It is also important to note that the use of these statistics assumes that the data were collected accurately through unbiased sampling procedures.

Table 7

Current Sample Sizes Used.

Most Common Sample Sizes Used for National and Regional Studies, by Subject Matter

Subject matter	National			Regional		
	Mode	$Q3$	$Q1$	Mode	$Q3$	$Q1$
Financial	1000+	—	—	100	400	50
Medical	1000+	1000+	500	1000+	1000+	250
Other behavior	1000+	—	—	700	1000	300
Attitudes	1000+	1000+	500	700	1000	400
Laboratory experiments	—	—	—	100	200	50

Typical Sample Sizes for Studies of Human and Institutional Populations

Number of subgroup analyses	People or households		Institutions	
	National	Regional or special	National	Regional or special
None or few	1000–1500	200–500	200–500	50–200
Average	1500–2500	500–1000	500–1000	200–500
Many	2500+	1000+	1000+	500+

Source: Seymour Sudman, *Applied Sampling*. New York: Academic Press, 1976, pp. 86–87.

Further Readings

Babbie, Earl R., *Survey Research Methods*. Belmont, California: Wadsworth, 1973.

Sudman, Seymour, *Applied Sampling*. New York: Academic Press, 1976.

Yamane, Taro, *Elementary Sampling Theory*. Englewood Cliffs, New Jersey: Prentice-Hall, Inc., 1967.

Part 5
How to Compare Options

Chapter 14
LOCATION QUOTIENT

Definition

The term *location quotient* can refer to either a calculated value or a technique. A location quotient is a simple ratio calculation that measures the concentration of some activity in a region being studied relative to a reference region. The technique is frequently used with economic data where it is a short-cut technique to determine where industries are concentrated and which economic sectors produce for the export market. Chapter 15 (Indices) and Chapter 10 (Multiplier Analysis), describe the uses of other indices and approaches that have similar uses.

An example using economic data could be an analysis to determine if employment in the electric and electronic equipment sector is more highly concentrated in the county of interest than in the state as a whole. Economic data are categorized by Standard Industrial Classification (SIC) code. The electric and electronic equipment sector is SIC 36. If it is more highly concentrated in an area, the assumption is made that some of the production is exported from that area.

Method

The location quotient is the division of one ratio by a second ratio. The first ratio calculates the percentage of total activity that the activity being studied represents in the study region. The second ratio calculates the percentage of total activity that the activity being studied represents in the reference region. The division of the first ratio by the second ratio produces a location quotient.

$$LQ = (E_{sj}/E_s) / (E_{rj}/E_r),$$

where

LQ = location quotient,
E = the measure of the activity being studied,
s = the specific region being studied,
j = the specific activity being studied, and
r = the reference region.

A location quotient of unity (one) indicates that the specific activity being studied is as concentrated in the region being studied as it is in the reference region. LQs less than one indicate the activity is less concentrated in the region being studied, and LQs greater than one indicate the activity is more concentrated in the region being studied than in the comparison region.

Example 1

This example results from a real study of 1980 employment data for selected counties in Wisconsin using the state of Wisconsin as the reference unit (see Table 1).

$$LQ_{\text{Dane, fabm}} = (E_{sj}/E_s) / (E_{rj}/E_r)$$
$$= (5{,}408/323{,}545)/(42{,}352/4{,}705{,}648) = 1.86.$$

In this example, Douglas County has few employees in the fabricated metals sector and a smaller proportion of its population employed in this sector than the average proportion of employment in fabricated metals in the state of Wisconsin. Dane and Milwaukee counties have a large number of employees working in this sector. Both counties have LQs of greater than 1, indicating a greater proportion of

Table 1

Concentration of Employment in Two Economic Sectors
in Selected Wisconsin Counties in 1980

Counties	Total Pop.	Fabricated Metals Mfg. SIC 34	LQ_{fabm}	Business Services SIC 73	LQ_{bserv}
Dane	323,545	5,408	1.86	1,002	0.11
Douglas	44,421	52	0.13	175	0.14
Milwaukee	964,988	20,186	2.32	22,082	0.84
Wisconsin	4,705,648	42,352		128,526	

Source: U.S. Department of Commerce, 1981.

their county populations are employed in fabricated metals production than the proportion of the state's population working in this sector. Because their location quotients are greater than 1, we conclude that Dane and Milwaukee counties export some of their fabricated metals production. In this instance, the assumption that they export is based on the fact that they both have a concentration of fabricated metals production that is greater than average state levels.

For those areas with LQs greater than one, the second part of the location quotient calculation estimates the quantity of exports. The data used can be employment, valued added, regional product, or other measures of economic activity.

$$_x\hat{E}_{sj} = E_{sj} - (E_s \times E_{rj}/E_r),$$

where:

$_x\hat{E}_{sj}$ = estimated export employment in the study region (s), for the specific activity (j) (note that this is equal to the difference between the observed levels of production and the levels that would have be expected if the subregion were perfectly in line with the reference region),

E_{sj} = specific employment activity in the study region,

E_s = total employment in the study region,

E_{rj} = specific employment activity in the reference region (often the nation), and

E_r = total employment in the reference region.

Employees producing fabricated metals products for export in Milwaukee County ($_xE_{sj}$) can be estimated using the location quotient formula:

$$\hat{_xE_{sj}} = \text{Milw. Cnty. Fabr. Metals Employment}$$
$$- \left(\text{Total Milw. Cnty. Pop.} \cdot \frac{\text{Wis. Fabr. Metals Employment}}{\text{Total Wis. Employment}} \right)$$
$$= 20{,}186 - (964{,}988 \times 42{,}352/4{,}705{,}648) = 11{,}501.$$

The interpretation is that the equivalent of 11,501 of the Milwaukee County employees working in the fabricated metals sector are producing fabricated metals products that will be used outside the county, i.e., exported. The other 8,685 employees working in the fabricated metals sector in Milwaukee (20,186 − 11,501) are assumed to be producing fabricated metals products that will be consumed in Milwaukee County.

Many variations of the location quotient calculations are available and will produce different results. In this example, we could have used the total number of employees instead of the total population for the counties and the state. We could have used a different reference region such as the United States, the Great Lakes region, or the 20 largest U.S. cities instead of using the State of Wisconsin. Each of these variations would have produced somewhat different results. Knowledge of the subject being analyzed and the purpose of the analysis should determine how the location quotient calculations are used. ■

Example 2

This example based on a published article, using 1980 census data for counties in Wisconsin and using the state of Wisconsin as the reference unit, illustrates the use of location quotient analysis to measure the concentration of segments of the population or activities (See Table 2.).

$$LQ_{\text{Dane, disabled}} = (E_{sj}/E_s) / (E_{rj}/E_r)$$
$$= (11{,}407/323{,}545)/(199{,}407/4{,}705{,}648) = 0.83.$$

In this example, Douglas and Milwaukee Counties have a higher concentration of disabled people than the state as a whole while Dane

Table 2

Concentration of Disabled People and Users of LP Gas
for Home Heating in Selected Wisconsin Counties

Counties	Total Pop.	Disabled	$LQ_{disabled}$	LP Gas	LQ_{lpgas}
Dane	323,545	11,407	0.83	6,370	0.72
Douglas	44,421	2,129	1.13	837	0.68
Milwaukee	964,988	45,698	1.12	2,626	0.02
Wisconsin	4,705,648	199,407		129,476	

Source: Page and Sawicki, 1984.

County has a smaller concentration. The very low LQ for Milwaukee County for LP gas indicates that a very small proportion of residents use this fuel for heating. ∎

Further Readings

Greytak, Edward, "A Statistical Analysis of Regional Export Estimating Techniques," *Journal of Regional Science*. **9**(3): 387–395, 1969.

Krueckeberg, Donald A., and Arthur L. Silvers, *Urban Planning Analysis: Methods and Models*. New York: John Wiley and Sons, 1974.

Page, G. William, and David S. Sawicki, "Teaching Computer and Policy Analysis Skills in a Case Study Course," *Journal of Planning Education and Research*. **4**(1): 43–54, 1984.

U.S. Department of Commerce, Bureau of the Census, *County Business Patterns, for the years* 1980 *and* 1985 *for Wisconsin and for the United States*. Washington, DC: Government Printing Office.

Chapter 15
INDICES

Definition

An *index* is a number or statistic that can be used to summarize or aggregate several measurements. "The concept *index* usually implies that the procedure used gives only an imperfect indicator of some underlying variable which is not directly measurable." (Blalock, 1979, Section 2.1). This implies that there is both an underlying variable and an indicator of this variable. Unfortunately, there is no logical method of determining whether a given operational definition (index) really measures the theoretically defined concept or variable. In practice, people simply agree that a given operational definition should be used to measure a certain concept if the operational definition seems reasonable on the basis of the theoretical definition. This may lead to several different indices being developed for the same variables, each producing different results (Allison, 1978). See Chapters 10 (Multiplier Analysis) and 14 (Location Quotient) for examples of specialized indices. For alternative methods for making decisions when multiple factors are involved, see Patton and Sawicki (1986, Chapter 8).

Method

There is no standard method. Creating an index is as much art as it is science. There are certain guidelines in the creation of indices. The most important is that you should not spend most of your time on the details of the quantification system at the risk of obtaining precise rankings of the wrong criteria.

An index is developed to combine into one variable several factors that measure some aspect of what one is trying to investigate. The factors that are combined into a single index value may be measured in different units and different scales of measurement. It simplifies the comparison of cases that involve multiple factors. One common example is a socioeconomic status (SES) index. An SES index is a single number combining income, education, race, and other factors.

The two problems that must be overcome are: (1) combining variables with dissimilar units or scales of measurement, and (2) achieving a meaningful measurement scale for the calculated index. Creating an index is like adding apples and oranges. Comparing the SES index for different places requires the combination of data on income, years of education, race, etc. The units and scales of measurement often are different. One useful approach is to to convert all of the variables into percentages before combining them in an index. For each city, county, or state, you could calculate the percentage above the median income, the percentage above the median years of education, the percentage non-white, or other measures of the factors you think are important. Without conversion to percentages, the size of some places could give them large index values even though they did not have a high concentration of the dimension the index was attempting to measure. Calculating location quotients for each of the potential variables allows a quick assessment of each variable's relative importance to each unit of analysis (see Chapter 14, Location Quotient).

Once the variables of interest are combined into a raw index value, what is the scale of that index? How much higher or lower is one number than another? Some form of standardized index is important. The z-score values associated with each raw index value are an effective standardized index, if the components being combined in the

Indices

index can be approximated by a normal distribution and are being combined as a weighted sum (see Chapter 12, Confidence Levels). Another popular standardization is to express the index in terms of one (or 100 percent). An index of 1.07 would be seven percent above the norm.

Using weightings in combining the factors into an index is problematic. There is often theoretical justification that one factor is more important in the index than others. In the United States, income is widely considered the most important factor in SES. The danger in using weightings is that they can dramatically change the index in ways that are not always intuitively clear. Using weightings that are percentages so that the sum of the weightings is one, or 100 percent, is useful to keep the systems of weightings comprehensible. Using multiplicative indices, indices that combine the factors by mathematical operations other than addition or by multivariate statistical procedures, may sometimes be appropriate, but the resulting index is complex and may be difficult to explain or defend.

Example 1

A recent study used an index to compare cities in terms of their number of new business starts. Such an index may be more useful than using raw numbers that don't take into account the size of the city or other factors. The index can be easily calculated by dividing the number of new business starts per 100,000 population in each city by the number per 100,000 in the country as a whole and then subtracting one. If a city had the same rate as the country, its index would be zero. In 1986, there were 253,092 new business starts in the U.S. The 1980 census count was 226,542,580.

Index values less than zero had fewer new business starts in 1986 per 100,000 population than the nation as a whole. The index created in Table 1 reveals that of the twenty largest cities in 1986, Detroit, New York City, and Philadelphia, are doing the worst in new business starts and that Dallas, Houston, and Cleveland, are doing the best. ■

Table 1
New Business Starts for the 20 Largest U.S. Cities in 1986

City	1980 Pop.	New Starts 1986	Starts per 100,000	City / Country	Index for 1986
New York	7,071,639	5,471	77.37	0.692	−0.31
Chicago	3,005,072	2,833	94.27	0.844	−0.16
Los Angeles	2,966,850	2,752	92.76	0.830	−0.17
Philadelphia	1,688,210	1,395	82.63	0.740	−0.26
Houston	1,595,138	3,834	240.36	2.151	1.15
Detroit	1,203,339	697	57.92	0.518	−0.48
Dallas	904,078	2,415	267.12	2.391	1.39
San Diego	875,538	1,164	132.95	1.190	0.19
Phoenix	789,704	1,341	169.81	1.520	0.52
Baltimore	786,775	1,331	169.17	1.514	0.51
San Antonio	785,880	1,406	178.91	1.601	0.60
Indianapolis	700,807	779	111.16	0.995	−0.01
San Francisco	678,974	1,045	153.91	1.378	0.38
Memphis	646,356	671	103.81	0.929	−0.07
Washington, DC	638,333	853	133.63	1.196	0.20
Milwaukee	636,212	665	104.52	0.936	−0.06
San Jose	629,442	715	113.59	1.017	0.02
Cleveland	573,822	1,360	237.01	2.121	1.12

Source: Dunn and Bradstreet, 1987.

Note: *Starts per* 100,000 *population* is calculated by dividing the number of new starts by (1980 population/100,000); New York = 5,471/70.71639 = 77.37. *City/country* is calculated by dividing starts per 100,000 by (253,092/2265.42580); New York = 77.37/(253,092/2265.42580) = 0.692.

Index for 1986 is calculated by subtracting one from city/county; New York = 0.692 − 1 = −0.31.

Example 2

This example deals with the creation of an index of need for home heating assistance funds for Wisconsin counties. Some combination of the following variables collected for each county could be combined into a single index: housing built before 1950, total population, and poverty population. The variables considered important in the index could then be combined to produce a raw index value. This raw value could be used, or it could be standardized to the normal distribution because the variables that are combined in the index are reasonably normal (see Chapter 1, Descriptive Statistics). The formula to standardize the index is

$$\text{standardized index} = (X - \overline{X})/S,$$

where:

X = the raw index value of each county,
$\overline{X}$ = the mean of the X values for all 72 counties in Wisconsin, and
S = the standard deviation of the X values for all 72 counties in Wisconsin (Page and Sawicki, 1984).

Table 2 presents data for two counties and the State of Wisconsin. Milwaukee County is the largest urban county in Wisconsin. Wood County is a rural county in central Wisconsin. The data are presented in original form and also in percentage form. The percentage form is each county's percentage of the state total.

Table 2

Indicators for Milwaukee and Wood Counties and Wisconsin

County	Total Population		Income Below Poverty Level		Housing Built before 1950	
Milwaukee	964,988	(20.5%)	18,664	(25.7%)	131,386	(23.8%)
Wood	72,799	(1.5%)	953	(1.3%)	8,007	(1.4%)
Wisconsin	4,704,963		72,594		553,204	

Source: U.S. Bureau of the Census, 1981.

An index for counties can be developed from either the raw numbers or the percentage data. A simple additive index is created by summing the three variables for each county. Using the raw numbers, the unstandardized index values are:

Milwaukee Cnty. = 964,988 + 18,664 + 131,386 = 1,115,038,

Wood Cnty. = 72,799 + 953 + 8,007 = 81,759.

Standardizing these numbers makes them a more useful index. The mean and the standard deviation needed to standardize the index are obtained by repeating the addition of the three variable for each of the 72 counties in Wisconsin and then calculating the mean and standard deviation of the resulting unstandardized index values for all 72 counties. The mean ($\overline{X}$) of all of the summed values for these three variables for all 72 counties in Wisconsin is 125,336, and the standard deviation (S) is 32,652.

The standardized index calculation for each county is:

$$\text{standardized index} = X - \frac{\overline{X}}{S},$$

where:

X = the unstandardized index value that is the sum of the three variables for each county.

Milwaukee Cnty. = $(1,115,038 - 125,336)/32,652 = 30.3$,

Wood Cnty. = $(81,759 - 125,336)/32,652 = -1.3$.

A standardized index of zero is average for an index standardized to the normal distribution. This standardized index immediately tells us that Milwaukee County (30.3) is far above average in Wisconsin in terms of this index measuring the need of counties for home heating assistance. Wood County (–1.3) is below average. The large population, large number of families with income below the poverty level, and the large quantity of old housing in Milwaukee County cause the standardized index for the county to be high. High index values would be interpreted to indicate a high need for heating assistance funds. ■

Example 3

This example deals with an index of technological innovation for countries. Data were collected over a number of years on patents for new technologies. How many times a patent was cited in the following years by new patent applications was used as the primary measurement of the index of national innovation. First, the patents of each nation were ranked according to how many times they were cited. Second, the top 10 percent of the most highly cited patents for all patent classes combined were categorized by the nationality of the patent holder. If the top 10 percent of patents held by a nation were among the top ten percent of all patents cited, that nation would have an index value of 1.0. A nation with an index value of 0.92 would have been eight percent below the expected index value of 1.0, which assumes all nations were equally innovative (Bond, 1987; Broad, 1988). Note that the number of tenths that this index differs from 1.0 is equivalent to the percentage difference because this index is standardized to 100 percent. Table 3 presents an example of such an index of technological innovation calculated for Japanese patents.

Table 3
Index of Japanese Technological Innovation

Year of Citation	# of Patients	% of All Patients in Top 10%	% of Japan Patients in Top 10%	Ratio of Japan to All Patients in Top 10%
1975	6371	13.5	16.4	1.2
1976	6567	12.8	16.4	1.3
1977	6220	11.3	15.6	1.4
1978	6668	12.8	16.7	1.3
1979	5279	10.0	14.0	1.4
1980	7160	16.6	21.8	1.3
1981	8425	11.3	16.6	1.5
1982	8177	13.9	19.1	1.4

Source: Narin and Olivastro, 1988.

The ratio of Japanese patents in the top 10 percent of total patents to the total patents in the top 10 percent is the index of technological innovation. For Japan during the years 1975 to 1982, this index ranges from 1.2 to 1.5. This index suggests that Japan ranged from being 20 to 50 percent more technologically innovative than the "average" industrial nation during this period.

See Chapter 14, on Location Quotient, for another application of an index standardized to 1.0. ■

Example 4

An example of a groundwater pollution potential index follows. This index is of interest because it demonstrates how a combination of disparate factors measured on different scales can be combined into a single index. The index is known as the "DRASTIC Index" (Aller *et al.*, 1985). The word DRASTIC is an acronym of the seven factors combined to create the single index value to assess the potential of a specific location, area, or region, to have its groundwater polluted.

The DRASTIC index is interesting because it combines some factors, for example, depth to water table, that are measured on an interval scale, with some factors, for example aquifer media, that are measured on a nominal scale. These dissimilar factors are combined in the DRASTIC Index by creating two ordinal scale measures for each factor, combining the two ordinal scales by means of multiplication, and then adding the resulting number for each factor. The two ordinal numbers are a "rating" and "weight." The following equation is used to calculate the DRASTIC Index:

POLLUTION POTENTIAL = $D_R D_W + R_R R_W + A_R A_W + S_R S_W + T_R T_W + I_R I_W + C_R C_W$,

where:

$D, R, A, S, T, I,$ and C are the seven factors (see Table 4),
R = rating, and
W = weight.

The seven factors combined to create the single value of the DRASTIC Index are presented in Table 4. Each of the seven factors has both a "rating" and a "weight." The weights were determined by a panel of experts according to their potential significance to pollution (see Table 4).

For a specific location, the ratings convert both interval scale data and nominal scale data into a single ordinal scale rating for each factor. Nominal scale factors, for example aquifer media, are assigned a rating according to the relative pollution potential of the various aquifer media. Interval scale factors, for example depth to water table, are categorized into "ranges," for example 50 to 75 feet, with a "rating" assigned to each of the ordinal "ranges" (see Table 5).

Table 4

Factors Comprising the DRASTIC Index and Their Weights

Symbol	Feature	Weight
D	Depth to water table	5
R	Net recharge	4
A	Aquifer media	3
S	Soil media	2
T	Topography	1
I	Impact of vadose zone	5
C	Hydraulic conductivity of the aquifer	3

Source: Aller *et al.*, 1985, p. 83.

Table 5
DRASTIC Index Calculation for Area with Glacial Till over Limestone

Feature	Range	Rating	Weight	Number
Depth to water table	30–50	5	5	25
Net recharge	4–7	6	4	24
Aquifer media	karst limestone	10	3	30
Soil media	clay loam	3	2	6
Topography	2–6%	9	1	9
Impact vadose zone	silt/clay	1	5	5
Hydraulic conductivity	2000+	10	3	30

DRASTIC Index =129

Source: Aller et al., 1985, p. 83.
Note: A vectorial approach is an alternative to using an index to compare the environmental conditions of sites (Halfon, 1989).

Further Readings

Aller, Linda, T. Bennett, Jay Lehr, and R. Petty, *DRASTIC: A Standardized System for Evaluating Ground Water Pollution Potential Using Hydrogeological Settings*. Robert S. Kerr Environmental Research Laboratory, ORD, U.S. Environmental Protection Agency, Ada, Oklahoma, EPA/600/018, 1985.

Allison, Paul D., "Measures of Inequality," *American Sociological Review*. **43**: 865–880, 1978.

Blalock, Hubert M., Jr., *Social Statistics, Revised Second Edition*. New York: McGraw-Hill Book Company, 1979.

Bond, J., *International Science and Technology Data Update 1987*. Washington, D.C.: National Science Foundation (NSF 87–319), 1987.

Broad, W. J., "Novel Technique Shows Japanese Outpace Americans in Innovation," *New York Times*, p. 1, March 7, 1988.

Dunn and Bradstreet Corporation, *Business Starts Record 1986/1987*. New York: The Dunn and Bradstreet Corporation, 1987.

Halfon, E., "Comparison of an Index Function and a Vectorial Approach Method for Ranking Waste Disposal Sites," *Environmental Science and Technology*. **23**(5): 600–609, 1989.

Narin, Francis, and Dominic Olivastro, "Identifying Areas of Leading Edge Japanese Science and Technology." *A Report to the National Science Foundation*. Haddon Heights, New Jersey: CHI Research/Computer Horizons, Inc., 1988.

Page, G. William, and David S. Sawicki, "Teaching Computer and Policy Analysis Skills in a Case Study Course," *Journal of Planning Education and Research*. **4**(1): 43–54, 1984.

Patton, Carl V., and David S. Sawicki, *Basic Methods of Policy Analysis and Planning*. Englewood Cliffs, New Jersey: Prentice-Hall, 1986.

U.S. Bureau of the Census, *Census of Population and Housing*. Washington, DC: U.S. Government Printing Office, 1981.

Chapter 16
NET BENEFIT EVALUATION

Definition

Net benefit evaluation describes a group of techniques that help us decide which alternative policy or project is optimal. In many situations we are called upon to select one policy, plan, or project from a set of possibilities. Quantitative methods are especially useful in deciding between alternative projects which differ in many important respects.

Different criteria exist to assist us in selecting the best alternative among a set of alternate projects. Decisions are often made for political reasons. Simple quantitative criteria such as maximizing benefits or minimizing cost are widely used. A variety of more sophisticated methods, including cost–effectiveness analysis, benefit–cost analysis, or internal rate of return analysis, often are used. Techniques are available to evaluate non-economic outcomes using goals achievement matrices and other methods. This chapter concentrates on using the *net present value* concept to evaluate alternatives. For most policy analysis applications, the net present value concept is the preferred evaluation method (Stokey and Zeckhauser, 1978).

Net present value, sometimes referred to as *net benefit,* is the sum of all discounted benefits and discounted costs for the life of the project. Discounting allows us to calculate the present value of future costs and benefits so that we can compare current costs and benefits to those we anticipate in the future. Discounting takes into account the opportunity cost, or the additional value of having an asset in the present instead of at some future date because we could invest that asset and earn interest or profits.

Method

The approach is to: (1) project the benefits and costs of each alternative venture into the future for the life of the venture, and (2) convert those summed benefits and costs into a common metric value (usually dollars) in the present so that each alternative venture can be compared. A negative net present value indicates that the costs of the venture exceed the benefits. In general terms, those ventures with the greatest net present value are recommended.

For each year of the venture, the discounted benefits and costs must be calculated. Net present value is calculated by summing the difference in the discounted benefits and the discounted costs for each year of the venture.

$$\text{NPV} = \sum_{i=1}^{n} \frac{B - C}{(1 + r)^n},$$

where

NPV = net present value,
Σ = the sum of the following term for each year from the first year ($i = 1$) to the last year (n^{th} year),
B = the benefits of the project measured in dollars in each year of the project,
C = the costs of the project measured in dollars in each year of the project,
r = the discount rate, and
n = the number of years of the project's benefits and costs.

Net Benefit Evaluation

Determining the discount rate is a critical step in performing discounting, in calculating net present value, or using other evaluation techniques (Baumol, 1970; Mishan, 1972). While many economic and political factors are important in evaluating projects, the selection of the discount rate can often determine the outcome. A funding agency or government policy often determines the discount rate. When the discount rate must be selected, it can become a political decision. Low discount rates increase the relationship of benefits to costs for projects over time, thus making projects more likely to be selected. High discount rates have the opposite effect.

The correct discount rate should be the real cost to society of borrowing money: the social opportunity cost. This is a theoretical rate that is usually approximated by one of two measures: (1) the interest rate government pays to borrow money, or (2) the after-tax opportunity cost of private capital.

Example 1

An agency purchases a computer workstation for $90,000. Let us assume that the one-time purchase price includes hardware, software, training, and maintenance for the year of purchase. Future years have a $2000 maintenance fee. For this example, let us assume this workstation will last only four years and that the use of this computer workstation produces $30,000 of benefits to the agency each year. The discount rate is determined to be seven percent. The data for this example and the calculations necessary for finding net present value are provided in Table 1.

The discounted benefits and costs are calculated for each year because the discount factor is different each year. The discounted benefits and costs are calculated separately in Table 1. The net present value row is calculated by subtracting discounted costs from discounted benefits. For example, in year number one, *net present value* is *discounted benefits* ($28,037) minus *discounted costs* ($1,869) equaling $26,168. Alternatively, the difference in benefits and costs for any year could be divided by the discount factor for that year to calculate that year's net present value. Again using year one as an example, the sum of *benefits* ($30,000) minus *costs* ($2000) divided by the *discount factor* (1.07) is $26,168.

Table 1

Net Present Value Calculations

	\multicolumn{6}{c}{Year}
	0	1	2	3	4	Total
Discount rate (r)	7%	7%	7%	7%	7%	7%
Discount Factor $(1+r)^n$	1	1.07	1.1449	1.2250	1.3108	
Benefits (B)	30,000	30,000	30,000	30,000	30,000	
Discounted benefits $[B/(1+r)^n]$	30,000	28,037	26,203	24,490	22,887	131,617
Costs (C)	90,000	2,000	2,000	2,000	2,000	
Discounted costs $[C/(1+r)^n]$	90,000	1,869	1,747	1,633	1,526	96,775
Net present value	−60,000	26,168	24,456	22,857	21,361	34,842

Notes: Computationally, it is convenient to calculate discounted benefits, discounted costs, and their difference for each year (the net present value for each year), and then to sum them. Benefits, costs, and net present value are measured in dollars. In year zero, the discount factor is one because any expression raised to the zero power is equal to one.

In this example, the net present value of the proposed purchase of a computer workstation has a positive value of $34,842. A positive dollar value indicates that the investment of the $90,000 could be a good investment. The significance of this technique is that it enables the analyst to compare this proposal for a computer workstation to other proposals for the use of agency resources. The other alternative projects may have different benefits and the benefits and costs may extend over a different number of years. By converting benefits and costs to a common metric (dollars) and using discounting to calculate the net present value of all of the alternatives, we can compare projects that are as different as apples and oranges.

The net present value concept produces the basic information that allows the analyst to compare alternative projects. Selecting the largest net present value (excess benefits) as the decision criterion favors large projects that produce large dollar values of discounted benefits in excess of discounted costs. Smaller projects that produce greater benefits per dollar of investment would be at a disadvantage if only net present value were considered. An alternative use of the net present value concept is to select the highest ratio of discounted benefits to discounted costs as the decision criteria. This approach is known as *benefit–cost analysis.* It favors the most efficient alternative project. Using the benefit–cost ratio approach allows a fairer evaluation of small scale projects with large scale projects in the competition to receive funding.

Many hand calculators, electronic spreadsheets, and other computer programs have built-in algorithms that do discounting automatically once you have entered the data and selected the discount rate. Printed tables are available that give: (a) the net present value of one dollar n years in the future for different discount rates, and (b) the net present value of one dollar each year for n years in the future for different discount rates (see Appendix 14). ∎

Potential Problems

All of the quantitative evaluation methods have some common problems. In using these methods, the analyst must be careful to consider problems of externalities, complementarities, incidence of benefits and costs, inflation, and measuring benefits. These potential problems are usually best handled by addressing them explicitly in the analysis. Even if they can not be included in dollar terms, they should be articulated for each alternative project so that the decision maker is aware of them when comparing the projects presented in the evaluation analysis.

Externalities are effects of a project that extend beyond the scope of the project you are evaluating and affect other projects or goals. Externalities are unintended and usually negative effects that are also called *spillovers*. Costs to other projects or goals should be included in the evaluation of your alternatives.

Complementarities exist when the alternate project being evaluated has beneficial effects on several goals or programs, not only on the one you are evaluating. Such an alternative may not be the optimal alternative for the goal being evaluated, but it may provide the greatest benefit to society because it provides benefits to several societal goals and programs.

The *incidence of benefits and costs* refers to who receives the benefits of the project and who pays the costs of the project. Those receiving the benefits and those paying the costs are often not the same. Concerns about the incidence of costs and benefits are often called *equity concerns*. They are often important considerations in evaluating policy.

Inflation is a potential concern whenever future benefits and costs are considered. The possibility of substantial changes in the rate of inflation over time may be a problem. Changes in inflation may have profound effects on the benefits and costs of projects that extend long into the future. To the extent that you can predict inflation, you should explicitly include it in the projection of annual costs and benefits.

Benefits are often more difficult to measure than costs. Costs of proposed projects are usually measured in dollars. Benefits often are difficult to convert into the common metric of dollars. Projects that reduce air pollution, improve the quality of life, or reduce mortality rates are difficult to convert into dollar values. To use quantitative evaluation methods, benefits should be converted into a common metric such as dollars.

Further Readings

Baumol, William J., "On the Discount Rate for Public Projects," in *Public Expenditures and Policy Analysis*, Robert H. Haveman and Julius Margolis (eds.). Chicago: Rand McNally College Publishing Company, 1970.

Krueckeberg, Donald A., and Arthur L. Silvers, *Urban Planning Analysis: Methods and Models*. New York: John Wiley and Sons, 1974.

Mishan, Edward J., *Economics for Social Decisions: Elements of Cost–Benefit Analysis*. New York: Praeger Publishers, 1972.

Patton, Carl V., and David S. Sawicki, *Basic Methods of Policy Analysis and Planning*. Englewood Cliffs, New Jersey: Prentice-Hall, 1986.

Stokey, Edith, and Richard Zeckhauser, *A Primer for Policy Analysis*. New York: W. W. Norton, 1978.

GLOSSARY

Abscissa: horizontal axis.

Alpha: the probability of making a Type I error.

Alpha Level: the level of statistical significance set by the analyst.

Alternative Hypothesis (H_1): states that there is a difference between two variables or between a sample value and the value in the population from which the sample is selected.

Annualizing Data: the process to calculate the annual rate of change that, when compounded over a series of years, will equal the rate obtained when we divide that ending year by the beginning year, assuming a constant rate of change.

Annual Net Benefits: yearly benefits minus yearly costs.

Bar Charts: figures that compare the differences among mutually exclusive categories of grouped data using bar segments of differing heights. The bars describing the data categories do not abut one another.

Benefit–Cost Ratio: the ratio of discounted benefits to discounted costs. Note that the alternative with the highest benefit–cost ratio does not necessarily have the highest net present value.

Beta: the probability of making a Type II error.

Bivariate Tabular Analysis: tabular analysis that involves only two variables.

Black Boxes: procedures in which the process of quantitative analysis is hidden.

Causality: determining which variable affects the other; i.e., the one which acts as the independent or predictor variable.

Chi Square (χ^2): a measure of significance used to determine if a relationship exists. It can be used to examine data displayed in tabular form, and it is handy for quickly reexamining data presented by others. Compares expected and observed cell frequencies to measure the extent to which the differences may have been due to chance.

Cluster Sampling: used when it would be difficult to compile a list of elements in the population. Elements are clustered or grouped together, and then selected clusters are sampled.

Coefficient of Variation: a measure of dispersion; the standard deviation divided by the mean.

Complementarities: exist when the alternate project being evaluated has beneficial effects on several goals or programs, not only on the one being evaluated.

Confidence Interval: in making an estimate, answers the question of how far from the true value we are willing to be at a stated level of probability. Are we willing to accept a value plus or minus $1,000 of the true mean with 95 percent confidence, or will we accept only a value plus or minus $5 of the true mean? The confidence interval is the set of acceptable hypotheses.

Confidence Level: often referred to as the significance level, answers the question of how often we are willing to be wrong in our estimates: 1 out of 100 times, 5 out of 100 times, 10 out of 100?

Constant Share Model: a simple ratio projection model. See ratio projection model.

Glossary

Consumer Price Index (CPI): a weighted average in which the weights for each of the 400 items used to develop the CPI are the percentages of total income spent on each item.

Correlation Coefficient: a numerical value that summarizes the strength of relationship between or among variables.

Cost–Benefit Analysis: a tool for measuring the relative economic efficiency of a range of alternatives. Uses the selection of the highest ratio of discounted benefits to discounted costs as the decision criterion.

Cost–Effectiveness Analysis: a tool for finding the alternative that accomplished the specified goal at the lowest cost. Differs from cost–benefit analysis, which may be used to compare alternatives that have very different goals.

Cost–Revenue Analysis: a tool for evaluating the profitability of a proposed action. Only monitarized revenues and costs to the entity undertaking the action are considered. Sometimes called a *fiscal impact analysis*.

Degrees of Freedom: the number of quantities that are unknown minus the number of independent equations linking these unknowns. For data in a contingency table, this boils down to finding out how many cells you must have filled in before you can determine all the cells.

Dependent and Independent Variables: the independent variable is the one we suspect affects the behavior of the other, or dependent, variable.

Descriptive Statistics: can be used to assemble, tabulate and summarize information about a topic. There are two types of descriptive statistics: *measures of central tendency* and *measures of dispersion*.

Differential Shift: measures the "competitive effect" in shift-share analysis. It tells us about the relative competitive advantages of the region being studied.

Direct Costs: resources that must be committed to implement a policy or program. This includes borrowing costs, one-time fixed costs, and operation and maintenance costs.

Direct Impact: an effect of a policy or program which addresses a stated objective of that policy or program.

Direct (Positive) Linear Relationship: a linear relationship in which the X variable as the independent variable predicts the value of the Y variable. As values of one variable increase, the values of the other variable also increase in a linear fashion.

Discounting: allows us to calculate the present value of future costs and benefits so that we can compare current costs and benefits to those we anticipate in the future.

Discount Rate: the rate estimated to approximate the time preference for money of the decision-making unit. Or, the rate at which a benefit declines in value if the decision-making body cannot have it now, but must postpone receiving it.

Disproportionate Sampling: used when we wish to ensure that a subpopulation contains enough cases for analysis. In order to assure a minimum number of respondents from a particular strata, we may take a larger percentage sample from that strata.

Dot Diagrams: are figures used when the variables have many categories or the data are grouped more finely than can be effectively illustrated with histograms.

ECF (Expected Cell Frequency): for the contingency table test

$$= \frac{\text{row marginal} \times \text{column marginal}}{\text{total number of observations}}.$$

Economic Efficiency: that allocation of resources that maximizes the welfare of society.

Economic Multipliers: mathematical functions used to project income or employment.

Economies and Diseconomies of Scale: the consequence that, as greater quantities of a commodity are produced, marginal costs of production may go up (*diseconomies*) or down (*economies*).

Effectiveness: an evaluation criterion. Measures whether the policy or program achieves its intended effect.

Glossary

Electronic Spreadsheet: computer software that permits the assembly of data into tabular format for ease of analysis.

Evaluation Methods: help decide which alternative policy or project is optimal.

Exponential Projection Model: a projection technique that may be appropriate when the rate of change consistently increases or decreases over the intervals of time in the historical data.

Externality: an effect, consequence, or phenomenon to which a free market assigns no value, either positive or negative, but which has a societal cost or benefit. Extends beyond the scope of the policy or project being studied. Externalities can be imposed on others by producers or consumers and can be positive or negative. Also called *spillovers*.

Extreme Values: a few values that are much larger or smaller than the rest of the data.

Fixed Costs: do not vary with the level of output.

Gamma or Yule's Q: a useful correlation measure when data can be structured in ordinal form. Ranges from -1.0 to ± 1.0, with 0.0 indicating no relationship between the variables, $+1.0$ indicating a perfect positive relationship, and -1.0 indicating a perfect negative relationship.

Grouped Data: data that have been converted from directly measured values into categories.

Histograms: figures that describe differences among categories of continuous grouped data. The bar segments describing the data categories abut one another.

Incidence of Benefits and Costs: refers to who receives the benefits of the project or program and who pays the costs of the project or program.

Independent and Dependent Variables: the *independent* variable is the one we suspect affects the behavior of the other, or *dependent*, variable.

Index: a number or statistic that can be used to summarize or aggregate several measurements. The concept of index usually implies that the procedure used gives only an imperfect indicator of some underlying variable that is not directly measurable.

Indirect Costs: the costs associated with impacts or consequences of a policy or program that are not directly related to that policy's or program's objectives.

Inferential Statistics: techniques used to make generalizations about a population from sample data.

Intangible Costs or Benefits: costs or benefits that cannot be measured in recognized units (pain and suffering, inconvenience, loss of confidence, etc.).

Interest Rate: the rate that the market will pay to have benefits now instead of later. Often the same as the discount rate, but not always.

Interval Scale Data: data classified on a scale that permits them to be measured exactly, using generally accepted units of measurement that can be infinitely divided.

Inverse (Negative) Relationship: as values of one variable increase, values of the other variable decrease.

Levels of Significance: guidelines that have been adopted for use in statistical analysis to aid decisionmaking. In social science research, the traditional significance level is 0.05. That is, results are considered statistically significant results if the values would occur by chance in no more than 5 out of 100 samples.

Linear Projection Model: appropriate for making projections when the subject has a history of nearly equal change for each time interval over the recent past.

Location Quotient: a simple ratio calculation that measures the concentration of some activity in a region being studied relative to a reference region.

Mean: the arithmetic average. Equals the sum of the values of the observations divided by the number of observations.

Glossary 201

Measures of Association: (such as gamma) tell us whether two or more variables are correlated and the strength of that correlation.

Measures of Dispersion: tell us how much the data deviate from the measures of central tendency.

Measures of Significance: (such as *chi square*) tell us the probability that an association occurred by chance.

Median: the measure of central tendency that identifies the mid-most value.

Mode: the measure of central tendency that is defined as the most frequently occurring value or category of the variable in the data.

Modified Exponential Projection Model: a projection method that may be appropriate where the rate of increase slowly declines as the subject being projected approaches some upper constraint on growth.

Multiplier: any constant term that is used in an arithmetic operation to estimate, apportion, or project some known quantity.

Multivariate Tables: have more than one independent variable used to explain the variation in the dependent variable.

Net Present Value: sometimes referred to as net benefit, is the sum of all discounted benefits and discounted costs for the life of the policy, program, or project.

Net Present Worth: see *present value*.

Nominal Scale Data: data classified into exhaustive, mutually exclusive, and named, but not ordered, categories.

Nonparametric Statistical Procedures: a wide range of statistical analyses that are comparable to those that could be accomplished with parametric statistical procedures if the data fit the assumptions required by parametric procedures.

Normal Curve: using the properties of the normal curve and the normal curve area table, we can answer a variety of questions: What is the probability that values above a given value of x will occur? What is the probability that values between x_a and x_b will occur? What percentage of the values lie above or below x_c? See normality.

Normal Curve Table: presents information about the way data are distributed within the normal curve and allows us to make estimates using z scores about the relationship among values we determine through sampling.

Normality: refers to the distribution of the data. If a simple frequency polygon of data has a classic bell-shaped curve, then the data are normally distributed. In the normal distribution, the mean, mode, and median are all the same value and are located in the center of the symmetrical distribution.

Null Hypothesis (H_o): states that there is no relationship between the two variables or there is no difference between the sample value and the value in the population from which the sample is selected.

Opportunity Costs: the value of resources diverted from or not available to other uses to make a given policy or program possible.

Ordinal Scale Data: data classified into exhaustive, mutually exclusive, and ordered or ranked categories.

Ordinate: vertical axis.

Outlier: an observation that comes from a different population than the rest of the observations in the data set (this implies that the outlier is really from a different distribution).

Parametric Statistical Procedures: powerful procedures for helping you analyze data and reach conclusions. Parametric statistical procedures require interval or ratio scale data that are normally distributed.

Pie Charts: circular figures that illustrate proportions or shares of the whole.

Population Pyramids: graphics used to show population distribution by sex and age for various points in time.

Present Value or Net Present Worth: the discounted future value (first discount costs and benefits separately, then find the net value of the two streams), using whatever discount rate has been determined appropriate in this case.

Projection Techniques: quantitative methods for estimating future conditions.

Glossary

Proportionality Shift: measures the "mix effect" in shift-share analysis. It tells us if the specific field being studied is a fast growth field, a slow growth field, or a declining field relative to total growth for all sectors of the economy at the reference region level.

p-value: a test of significance that allows us to calculate the extent to which the null hypothesis is supported by the data.

Quick Analysis: intended to be used as a first approximation that can be followed up on with more sophisticated techniques if time permits.

Quick Quantitative Method: a group of techniques that can be used as a first approximation to a more complex analysis, as means of a complete analysis by itself, or as a check on the reasonableness of an analysis.

Randomness and Representativeness: all members of the group being surveyed must have a known chance of being selected in the sample.

Random Sample: a scientific, unbiased sampling procedure which assures that all people in the population under study have an equal chance of being selected.

Range: the difference between the smallest and the largest values of the variable in the distribution (data set).

Ratio Projection Model: used when you decide to base your projection on an already prepared projection for some other entity. Usually occurs when trying to make a projection for a smaller entity, and when a projection is available for a larger entity of which the smaller entity is a part.

Ratio Scale Data: data classified on a scale that permits them to be measured exactly using generally accepted units of measurement, and which includes a nonarbitrary zero point.

Representativeness: See randomness.

Sample Mean: X with a bar above it ($\bar{X}$), pronounced: "X bar." The mean of a set of sample data.

Sample Size: the size of the sample taken affects the extent to which we can place confidence in the data collected through a survey.

Scales of Measurement: there are four scales of measurement for quantitative data: nominal, ordinal, interval, and ratio.

Scatterplots: graphic representations of two variables: one is measured on the *y* (vertical) axis and the second variable is measured on the *x* (horizontal) axis. Scatterplots, an extension of the concept of dot diagrams, are also called *scattergrams* or s*catter diagrams.*

SES (Socioeconomic Status) Index: a single number combining measures of income, education, race, and other factors.

Shift-Share Analysis: used to describe how the change in activity in one region is different from some reference region.

SIC Code: Standard Industrial Classification code. Used to classify economic activities.

Simple Random Sampling: involves selecting elements or members from a group or population at random. Usually a random number table is used to select a given number or percentage of elements from a list.

Standard Deviation: a measure of dispersion of values in relation to the mean of a distribution. Calculated by taking the square root of the variance.

Standard Error: measures the error of the sample statistic as an estimate of the population parameter.

Statistical Sampling: the process by which a portion of the whole (population or universe) is selected for examination with the intent to generalize from that sample to the entire population.

Statistical Significance: a measure of how likely it is that relationships or associations found in sample data describe characteristics in the population from which the sample was taken.

Step-Down Model or Constant Share Model: a simple ratio projection model. See ratio projection model.

Stratified Sampling: a sampling technique that helps assure we obtain a representative sample by breaking the population into homogeneous subsets that are then sampled.

Systematic Sampling: a method for selecting elements from a list to produce a random sample.

Glossary

Tabular Analysis: involves displaying data in a logical, consistent format in tables that permit easy and accurate interpretation.

Time-series Plots: these diagrams show change over time for a variable.

Trend Extrapolation Projection Models: make projections into the future based on trends in historical data.

Type I Error: occurs when we reject the null hypothesis when it is true.

Type II Error: occurs when we fail to reject the null hypothesis when it is false.

Variance: the arithmetic average of the squared deviations of values from their mean. The symbols for variance are: S^2 for sample data, and σ^2 for population (universe) data.

Weighting: a procedure for applying a factor or weight to the results of a sample to adjust for a disproportionate sample, to correct for a misestimate of the size of a cluster or strata, to adjust for different response rates, or to make population estimates.

$\bar{X}$: the symbol for the sample mean. The symbol for the population mean is μ.

$\tilde{X}$: the symbol for the median.

z Scores: allow us to measure in a standardized way, in the number of standard deviation units, how far any given value lies from the mean value.

APPENDICES

Appendix 1
A Checklist for Using *Quick Answers to Quantitative Problems*

The following checklist is presented as a suggested guide to the use of *Quick Answers to Quantitative Problems*. Experienced analysts find these or equivalent steps useful in approaching quantitative problems and in evaluating their answers. The steps focus on finding quick answers to questions that involve quantitative information. The checklist does not encompass the broader issues of policy analysis, such as formulating alternative policies, implementation, and evaluation.

1. Be sure that you understand what type of an answer you are seeking. Issues are often complex and it is easy to find yourself solving the wrong problem or only a small part of the problem. Take the time to read about the situation carefully and ask questions before starting to analyze the problem.

2. Attempt to find a previous analysis of the problem or of a closely related problem. Finding such earlier work is always informative and useful.

3. Assemble and organize all of the available data. It is usually best to do this at the beginning of the analysis.

4. Identify the chapter in *Quick Answers to Quantitative Problems* that describes the appropriate method for analyzing your problem. Start with the table of contents. Look through the chapters until you find the technique that you need.

5. Be sure that you understand the chapter in *Quick Answers to Quantitative Problems* that you have selected. Work through the example or examples that are closest to your problem. Be sure that the logic of the technique makes sense to you. Make sure that the interpretation of the results of using the techniques in the chapter examples makes

sense. Refer to the further readings provided at the end of each chapter if you have unanswered questions.

6. Evaluate the data for its pertinence to the problem and its quality. Do the data bear on the problem? Will the data allow you to reach conclusions about the problem? Are there missing data, data collection or other problems with the data that suggest that it may be inadequate for your purposes? Are additional data needed?

7. Do the calculations or techniques described in *Quick Answers to Quantitative Problems*. Redo the calculations to be sure that you did not make a simple arithmetic error.

8. Evaluate your results critically. Do your results and conclusions seem reasonable? If your results are counter-intuitive, they may be correct, but they should be subjected to extreme scrutiny to be certain that you have not made a mistake either in conceptualizing the problem or in conducting the analysis.

9. Evaluate whether the results of your quick analysis using *Quick Answers to Quantitative Problems* has produced sufficiently compelling information, or if a more thorough and time consuming analysis is required to resolve the problem at hand.

10. Ask a knowledgeable colleague or friend to review your work or listen to you present your results. If your results and conclusions are not persuasive to this person, you should reevaluate your work carefully. Start again at the first step of this checklist and work through the problem from the beginning.

11. If there are still questions in your mind about the technique you have used, your results, or your conclusion, seek out assistance. When you are unsure of your results, it is always better to seek assistance before you risk your reputation for professional competence.

Appendix 2
Words Commonly Confused

accept —to take
except—to exclude

ad—advertisement
add—to increase

adapt—to make fit
adopt—to take as one's own

adverse —unfavorable
averse—disinclined

advice—recommendation (noun)
advise—to recommend (verb)

affect —to influence
effect—to accomplish; result

aid—help
aide—assistant

alley—narrow back street
ally—confederate

all ready—completely ready
already —previously; so soon

altar—platform in a church
alter—to change

ambiguous—having several meanings
ambivalent—uncertain, of two minds

amend—to modify
emend—to edit or correct

ante- —before
anti- —against

appraise—to estimate the value of
apprise—to notify

assay—to evaluate
essay—composition; to try

auger—tool for boring holes
augur—to predict

avoid—to keep away from
evade—to dodge

bazaar—marketplace, fair
bizarre—strange

bloc—political grouping
block—obstruction

born—given birth
borne —carried, supported, produced, etc.

bough—tree branch
bow—to bend or yield

brake—to slow and stop
break—to fracture, damage, etc; to stop work temporarily

breach—a break
breech—the buttocks

breadth—expanse
breath—air inhaled and exhaled (noun)
breathe—to inhale and exhale (verb)

calendar—chart for reckoning time
calender—machine for processing paper, fabric, etc.
colander—device for draining food

callous—unfeeling
callus—hardened skin

Calvary—site of Jesus' crucifixion
cavalry—mounted soldiers

cannon—gun
canon—law

canvas—cloth
canvass—to solicit opinions, votes, etc.

capital —economic resources; government seat
capitol—legislature building

censer—container for incense
censor—one who checks for objectionable material

cession—act of ceding
session—meeting

chafe—to rub
chaff—worthless matter

chord—musical tones
cord—thin rope

cite—to quote
sight—power of seeing
site—place

climactic—referring to a climax
climatic—referring to a climate

complement —something that completes or balances
compliment—praise

compose—to make up
comprise —to include or consist of

consul—diplomat
council —assembly
counsel—advice; a lawyer

corporal—of the body
corporeal—material, tangible

corps—group of people
corpse—dead body

credible—believable
creditable—praiseworthy

cue—hint; rod used in billiards
queue—line

dairy—milk farm
diary—daily record book

delusion—false belief
illusion—false impression

desert—arid region; to leave or abandon
dessert—final course of a meal

diagnosis—medical analysis
prognosis—medical prediction

dialectal —of a dialect
dialectic—of logical argumentation

discomfit—to confuse, frustrate
discomfort—to make uncomfortable

discreet—circumspect, prudent
discrete—separate

dispute—to argue
refute—to disprove

dual—of two, double
duel—fight

emigrate—to leave a country
immigrate—to enter and settle in a country

eminent—renowned
immanent—inherent
imminent—about to happen

energize—to give energy to
enervate—to deprive of strength or force

Source: Random House, Inc. *The Random House Dictionary of the English Language,* Second Edition, Unabridged, 1987, pp. 2467–77.

Words Commonly Confused

equable—uniform
equitable—fair, just

faze—to disconcert
phase—stage or aspect

flair—aptitude, style
flare—to burn, burst out

flaunt —to make boastful display
flout—to treat with contempt

flounder—to struggle awkwardly
founder—to sink, fail

forceful—powerful
forcible—done by force

foreword—introduction to written work
forward—onward, ahead

fortuitous —happening by chance
fortunate—lucky

full—filled
fulsome—offensive because excessive

gamble—to bet
gambol—to frolic

hangar—shed for airplanes
hanger—frame for hanging clothes

hyper-—excessive, above
hypo-—insufficient, under

idle—inactive
idol—image of a god

incredible—extraordinary, unbelievable
incredulous—skeptical

indict—to charge with an offense
indite—to compose or write

inter-—between, among
intra-—within

its—belonging to it
it's—it is

jibe—to agree
jive—to fool

lay —to place or put; past tense of *lie*
lie—to recline

lead—a metal; to guide
led—past tense of *lead*

lesser—comparative of *little*
lessor—one who grants a lease

loath—reluctant
loathe—to hate

loose—not tight or bound; to make loose
lose—to experience loss

luxuriant—abundant, lush
luxurious—sumptuous

material—substance
matériel—equipment, arms

mean—intermediate value of number sequence
median—middle number in number sequence

meddle—to interfere
metal—a substance
mettle—spirit

militate—to have an effect
mitigate —to moderate, soften

miner—one who mines
minor—underage person

moral—ethical; lesson
morale—spirit

mucous—of a viscous substance (adjective)
mucus—a viscous substance (noun)

naval—of the navy
navel—umbilicus

ordinance—law
ordnance—military supply

palate—roof of mouth; taste
palette—artist's board
pallet—crude bed; platform

passed—past tense of *pass*
past—former time

peace—calmness; lack of hostility
piece—a part

pedal—foot lever
peddle—to sell

persecute—to hound
prosecute—to institute legal proceedings against

perspective—vision, view
prospective—future

plain—simple
plane—airplane; to smooth

practicable—feasible
practical—suited to actual conditions; sensible

precede—to go before
proceed—to continue

prescribe—to recommend
proscribe—to prohibit

principal —chief; head person; capital sum
principle—rule

prophecy—prediction (noun)
prophesy—to predict (verb)

prostate—gland
prostrate—lying flat

quiet—still
quite—very

role—a part
roll—to turn; a small bread

seasonable—appropriate to the season; timely
seasonal —depending on the season

shear—to clip
sheer—transparent; utter

stationary—fixed
stationery—paper supplies

than—as in "greater than"
then—at that time

their—belonging to them
there—at that place
they're—they are

to—toward
too—also; excessive
two—number

trail—path
trial—judicial proceedings

trooper—soldier or police officer
trouper—actor; dependable person

venal—corrupt
venial—pardonable

weather—state of atmosphere
whether—if

who's—who is
whose—belonging to whom

your—belonging to you
you're—you are

Appendix 3
Perpetual Calendars

Source: Ameritech Publishing, Inc., *Pages Plus*, Toledo, 1988–89.

Perpetual Calendars

213

Appendix 4
World Map with Time Zones

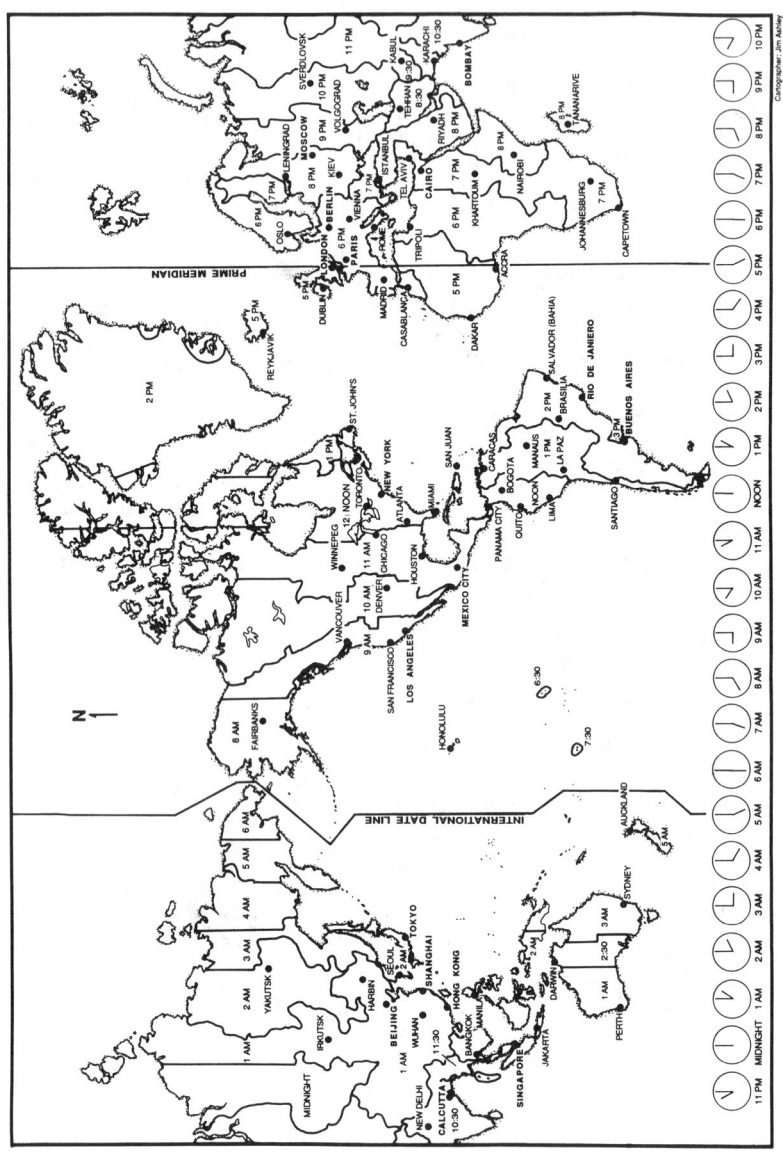

Appendix 5
United States Map with Time Zones and Telephone Area Codes

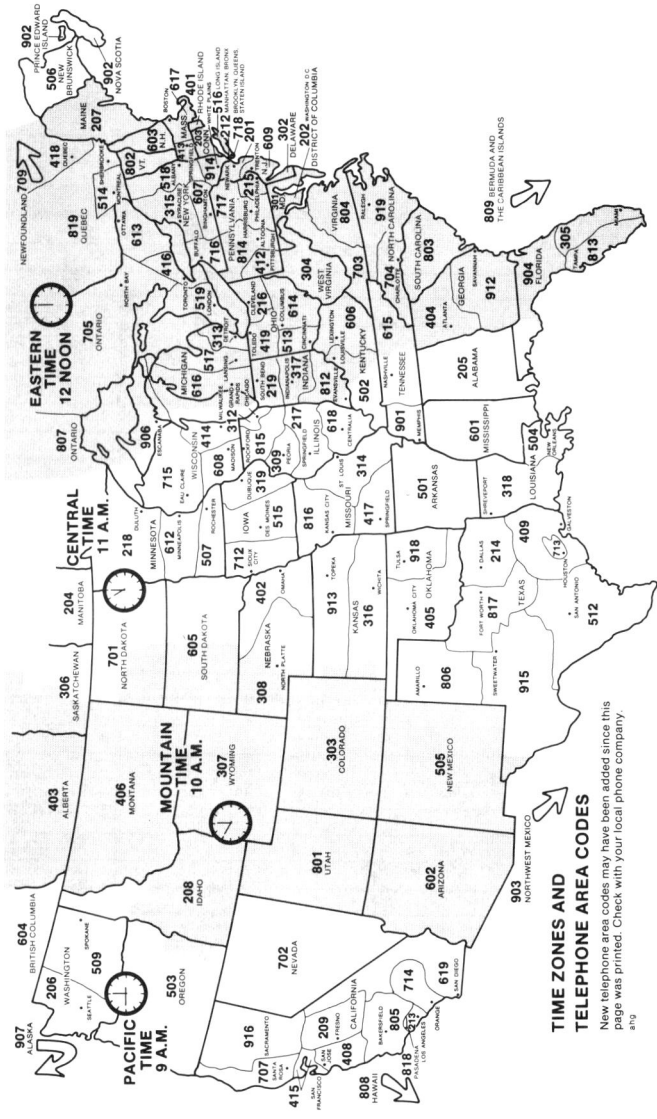

Source: Keith Clark Co., *Week at a Glance*, Professional Appointments, 1989.

Appendix 6

International Distances Chart

		ABU DHABI	5167	3260	14244	4975	5142	5972	4637	2003	10735	5158	13534	2367	3471	7498	11688	4845	2317	4903	4892	6071	13865	2987	2043	5478	13481
			AMSTERDAM	2164	18728	9185	1237	7841	577	6864	5575	174	11424	3282	7620	9647	6628	623	6368	690	367	9300	11676	2213	3350	359	8963
				ATHENS	16775	7933	1822	7633	1803	5179	7639	2092	11677	1120	6325	7979	8765	2136	5019	1710	2026	8560	13439	562	1256	2394	11118
					AUCKLAND	9566	19204	10388	17743	12294	14478	18279	10372	16573	11176	11796	13181	17525	12482	18609	17813	9121	7052	17042	16287	18330	10479
3211						BANGKOK	9692	3291	8613	3010	13733	9263	16885	7279	1610	10144	13789	8628	2917	9249	8824	1723	10634	7477	6895	9544	13319
2026	1345						BARCELONA	8822	1500	7044	5881	1063	10447	2897	8084	8502	7101	1760	6782	624	1473	10087	12766	2238	3122	1138	9677
8851	11637	10424						BEIJING	7375	4760	10860	7983	19265	7557	3271	12947	10626	7218	3788	8223	7492	1972	8171	7072	7135	8160	10082
3091	5707	4930	5944						BERLIN	6298	6098	654	11890	2891	7045	9588	7103	355	5791	876	255	8770	11782	1739	2903	934	9332
3195	769	1132	11933	6023						BOMBAY	12275	6891	14937	4363	1664	8216	12976	6430	1156	6725	6544	4311	12928	4818	4017	7205	14021
3711	4872	4743	6455	2045	5482						BOSTON	5598	8619	8737	12517	12411	1369	5904	11504	5929	5843	12831	8191	7783	8884	5280	4179
2881	358	1120	11025	5352	932	4583						BRUSSELS	11282	3212	7689	9490	6679	769	6427	533	491	9416	11825	2185	3302	320	9055
1245	4265	3218	7639	1870	4377	2958	3914						BUENOS AIRES	11811	16535	6891	8978	12046	15800	11045	11773	18463	12160	12235	12236	11105	9828
6671	3464	4747	8996	8534	3654	6748	3789	7628						CAIRO	5708	7208	9881	3206	4436	2816	3125	8158	14239	1234	426	3513	12223
3205	108	1300	11358	5756	661	4961	406	4282	3479						CALCUTTA	9684	12861	7083	1307	7651	7264	2654	11357	5867	5314	7978	13141
8410	7099	7256	6445	10492	6492	11971	7388	9282	5356	7011						CAPE TOWN	13658	9942	9284	8958	9725	11867	18562	8367	7481	9635	16054
1471	2039	696	10298	4523	1800	4696	1796	2711	5429	1996	7339						CHICAGO	6860	12047	7069	6850	12560	6849	8834	9978	6371	2810
2157	4735	3930	6945	1000	5023	2033	4378	1034	7778	4778	10275	3547						COPENHAGEN	5857	1145	289	8688	11428	2021	3191	958	9026
4659	5995	4958	7330	6303	5283	8045	5958	5105	7712	5897	4282	4479	6018						DELHI	6363	6020	3770	11930	4560	4032	6724	12882
7263	4119	5447	8191	8568	4413	6603	4414	8063	851	4150	5579	6140	7992	8487						GENEVA	862	9544	12358	1921	2959	748	9519
3011	387	1327	10890	5361	1094	4485	221	3996	3669	478	7485	1992	4401	6178	4263						HAMBURG	8934	11629	1988	3150	723	9098
1440	3957	3119	7756	1813	4214	2354	3599	718	7148	3994	9818	2757	812	5769	7486	3640						HONG KONG	8945	8034	7740	9646	11674
3047	429	1063	11563	5747	388	5110	544	4179	3684	331	6863	1750	4754	5566	4393	712	3954						HONOLULU	13068	13969	11653	4125
3040	228	1259	11069	5483	915	4655	159	4066	3631	305	7291	1942	4514	6043	4257	180	3741	536						ISTANBUL	1170	2504	11043
3772	5779	5319	5668	1071	6268	1225	5450	2679	7973	5851	11473	5069	1649	7374	7805	5399	2343	5931	5552						JERUSALEM	3615	12210
8616	7255	8351	4382	6608	7933	5077	7321	8033	5090	7348	7556	8848	7057	11534	4256	7101	7413	7679	7226	5558						LONDON	8778
1856	1375	349	10590	4646	1391	4394	1081	2994	4836	1358	7603	767	3646	5199	5489	1256	2834	1194	1235	4992	8120						LOS ANGELES
1270	2082	781	10121	4285	1940	4434	1804	2496	5520	2052	7603	265	3302	4649	6200	1983	2505	1839	1957	4810	8680	727					
3404	223	1488	11390	5931	707	5071	580	4477	3281	199	6901	2183	4957	5987	3959	595	4178	465	449	5994	7241	1556	2246				
8377	5570	6909	6512	8276	6013	6265	5799	8713	2597	5627	6107	7595	8166	9976	1746	5609	8005	5915	5653	7254	2563	6862	7587	5455			
3500	921	1475	12174	6336	314	5744	1163	4688	3410	818	6229	2085	5337	5304	4191	1289	4529	637	1111	6562	7874	1705	2238	785	5833		
7263	10280	9289	1634	4573	10458	5650	9924	6096	10521	10325	7226	8678	5547	6424	9673	9930	6333	10271	10057	4593	5507	9090	8521	10503	7930		
8932	5739	7024	6802	9739	5909	7754	6056	9739	2279	5757	4577	7700	9504	8515	1691	5921	9121	5962	5898	8796	3789	7114	7800	5560	1549		
2893	514	910	11482	5614	452	5031	523	4029	3838	434	6943	1599	4619	5493	4542	720	3816	155	560	5823	7764	1041	1685	597	6051		
6616	3428	4737	8935	8337	3677	6518	3740	7522	251	3451	5593	5427	7615	7919	744	3606	7013	3677	3581	7744	4918	4803	5502	3256	2469		
2321	1337	1386	10063	4393	1873	3610	1002	3129	4498	1404	8365	1801	3443	6277	4984	971	2702	1504	1109	4672	7048	1091	1660	1557	6085		
2126	4133	2828	8678	4485	3652	5727	3948	2816	7190	4066	6472	2186	3839	2542	8010	4155	3373	3764	4080	5449	10741	2952	2276	4229	9664		
6860	3654	4937	8816	8668	3842	6843	3979	7808	191	3669	5276	5618	7936	7799	713	3857	7319	3874	3820	8068	4969	5026	5711	3471	2451		
4808	5742	5797	5532	2615	6421	1110	5494	3956	6871	5840	9786	5796	2955	8906	6500	5363	3415	6036	5541	1549	4104	5448	5535	5919	5724		
6692	3512	4825	8835	8353	3772	6509	3820	7582	313	3536	5612	5516	7649	8010	645	3682	7061	3766	3661	7734	4819	4887	5588	3342	2366		
3260	266	1306	11521	5877	517	5118	547	4365	3446	163	6853	1998	4892	5783	4143	639	4102	257	464	5996	7449	1405	2075	212	5658		
5614	8779	7628	3312	3301	8788	4944	8427	4514	11621	8793	7839	6992	4163	5416	10979	8499	4877	8660	8579	3728	6777	7467	6850	8989	9337		
7310	5937	6033	7636	9993	5294	10766	6207	8338	4829	5844	1223	6141	9372	3775	5284	6321	8749	5673	6147	11005	8291	6380	6405	5750	6294		
2674	804	655	11433	5494	534	5061	735	3845	4102	729	6919	1327	4495	5230	4821	951	3684	433	813	5779	8038	857	1435	891	6346		
8145	5468	6792	6517	7930	5963	5918	567	8405	2699	5532	6453	7466	7828	10245	1859	5474	7693	5833	5533	6910	4261	672	7436	5369	347		
9097	7452	7797	6021	10968	6923	11842	7772	9984	5217	7375	705	7954	10961	4946	5294	7835	10518	7274	7677	11607	6861	8136	2005	7240	5578		
7527	6077	6221	7483	10196	5452	10933	6356	8558	4795	5985	1044	6345	9592	3949	5209	6462	8967	5826	6290	11221	8090	6567	6610	5885	6149		
4299	5332	5305	5963	2312	5982	595	5064	3488	6815	5425	12073	5284	2514	8519	6546	4948	2920	5601	5123	1303	4549	4956	5023	5519	5968		
4068	5530	5318	5815	1784	6119	669	5233	3131	7314	5616	12190	5199	2113	8053	7081	5143	2640	5753	5310	755	4955	4973	4934	5731	6507		
3669	6526	5629	5227	887	6767	2775	6169	2428	9410	6566	9873	5139	1794	6009	9375	6195	2574	6525	6306	1600	6726	5376	4924	6748	8784		
2978	701	1497	10565	5143	1417	4179	505	3878	3753	799	7793	2115	4204	6421	4286	325	3467	1032	505	5122	6872	1352	2064	892	5531		
6085	2875	9523	1343	4675	10677	5545	9998	6305	10092	10404	7345	8957	5668	6856	9242	9963	6472	10422	10111	4566	5065	9286	8778	10557	7497		
5018	5788	5922	5475	2865	6487	1307	5556	4195	6718	5888	11412	5957	3200	9157	6311	5415	3640	6101	5594	1798	3858	5574	5699	5956	5486		
6905	3728	5044	8624	8480	3989	6594	4035	7777	431	3754	5545	5734	7810	8134	437	3896	7243	3985	3877	7815	4659	5103	5806	3560	2176		
2635	582	797	11094	5251	1026	4647	326	3721	4045	570	7328	1480	4262	5653	4698	541	3465	500	462	5437	7634	794	1504	769	6116		
7063	3858	5141	8621	8806	4044	6941	4181	8002	395	3873	5194	5822	8102	7892	595	4058	7501	4075	4023	8163	4838	5231	5915	3676	2300		

MILES

Source: The Times Family Atlas of the World, First Edition, Topsfield, MA: Salem House, 1988.

KILOMETRES

Madrid	Melbourne	Mexico City	Milan	Montreal	Moscow	Nairobi	New York	Osaka	Ottawa	Paris	Perth	Rio de Janeiro	Rome	San Francisco	Santiago	Sao Paulo	Seoul	Shanghai	Singapore	Stockholm	Sydney	Tokyo	Toronto	Vienna	Washington
5632	11689	14374	4655	10647	3735	3422	11040	7737	10770	5247	9035	11764	4303	13107	14639	12113	6918	6547	5905	4792	9793	8075	11112	4240	11366
1482	16544	9236	827	5516	2152	6651	5881	9241	5651	428	14128	9555	1294	8799	11992	9779	8581	8899	10503	1128	4627	9315	6000	936	6209
2374	14949	11304	1464	7623	2230	4551	7945	9329	7765	2101	12276	9709	1054	10930	12548	10012	8537	8558	9058	2409	15325	9531	8117	1282	8274
19592	2629	10947	18478	14379	16194	13966	14187	8903	14218	18540	5361	12288	18400	10487	9690	12042	9596	9358	8411	17002	2161	8811	13879	17853	13873
10196	7359	15760	9035	13417	7070	7218	13949	4208	13442	9457	5313	16081	8842	12762	17651	16409	3720	2871	1427	8276	7523	4610	13647	8451	14172
506	16830	9510	727	5918	3014	5877	6183	10333	6071	832	14143	8520	860	9596	11141	8774	9626	9848	10890	2280	17182	10440	6420	1351	6508
9243	9093	12478	8097	10490	5809	9216	11012	1787	10475	8236	7957	17325	8144	9524	19057	17595	958	1077	4465	6725	8923	2104	10611	7479	11170
1871	15970	9746	842	6018	1612	6353	6403	8841	6147	880	13561	9989	1182	9129	12507	10229	8150	8421	9927	812	16090	8942	6494	524	6729
7545	9810	15673	6484	12105	5036	4531	12566	6367	12201	7024	7264	13419	6187	13526	16068	13772	5614	5038	3908	6241	10147	6751	12516	5988	12877
5487	16931	3668	6176	404	7239	11571	307	11058	504	5545	18701	7771	6602	4344	8395	7716	10967	11770	15144	6040	16241	10812	694	6509	635
1316	16616	9264	698	5553	2260	6544	5904	9398	5691	262	14150	9404	1173	8902	11868	9632	8731	9037	10567	1285	16743	9476	6041	918	6233
10024	11629	7366	11173	9001	13461	10416	8490	18749	9031	11029	12616	1968	11135	10384	1135	1680	19429	19618	15889	12541	11821	18365	8924	11793	8359
3355	13966	12392	2573	8733	2899	3518	9042	9327	8877	3215	11252	9882	2135	12015	12800	10211	8504	8367	8270	3404	14415	9587	9228	2382	9370
8589	8927	15295	7433	12254	5541	6178	12771	4756	12310	7873	6700	15083	7233	12598	17640	15436	4045	3401	2887	6765	9122	5150	12569	6858	13038
8536	10338	13703	8840	12744	10101	4090	12551	14332	12891	9307	8716	6075	8417	16487	7959	6355	13710	12959	9671	10334	11034	14737	13090	9098	12700
6744	15567	2721	7309	1198	8021	12890	1147	10460	1038	6667	17668	8504	7759	2992	8520	8383	10534	11395	15087	6898	14874	10157	703	7561	958
2075	15891	9529	1159	5803	1563	6687	6207	8631	5926	1029	13678	10173	1531	8809	12609	10400	7962	8276	9970	523	16034	8714	6269	870	6531
7288	10192	14679	6141	11286	4349	5428	11779	5496	11364	6601	7849	14080	5929	12380	16926	14430	4699	4249	4142	5579	10415	5857	11656	5576	12071
1025	16529	9594	249	5917	2421	6058	6235	9714	6061	414	13937	9129	696	9387	11706	9376	9014	9258	10500	1661	16772	9818	6413	804	6564
1788	16184	9491	901	5763	1785	6566	6147	8917	5892	747	13806	9892	1308	8904	12355	10122	8244	8546	10149	812	16272	9002	6239	744	6474
10561	7392	14155	9371	12462	7158	8769	12984	2493	12446	9650	5999	17710	9300	11121	18679	18058	2097	1215	2575	8243	7348	2893	12577	8749	13137
12671	8862	6098	12495	7915	11342	17286	7996	6604	7755	11988	10906	13343	12936	3857	11041	13020	7321	7974	10824	11059	8151	6208	7498	12285	7785
2744	14628	11448	1675	7730	1755	4751	8089	8767	7864	2261	12017	10268	1379	10809	13094	10568	7975	8003	8652	2175	14944	8970	8212	1277	8418
3602	13713	12552	2711	8854	2671	3662	9190	8907	8993	3339	11024	10308	2310	11966	13226	10637	8083	7941	7924	3322	14126	9171	9343	2421	9519
1264	16902	8947	960	5240	2506	6805	5586	9526	5379	341	14466	9254	1434	8640	11651	9470	8882	9223	10860	1436	16990	9585	5729	1238	5915
9387	12762	2492	9738	3973	9793	15553	3945	9211	3808	9106	15026	10129	10212	558	8976	9896	9605	10471	14136	8901	12065	8828	3502	9842	3701

	Melbourne	Mexico City	Milan	Montreal	Moscow	Nairobi	New York	Osaka	Ottawa	Paris	Perth	Rio de Janeiro	Rome	San Francisco	Santiago	Sao Paulo	Seoul	Shanghai	Singapore	Stockholm	Sydney	Tokyo	Toronto	Vienna	Washington
Madrid	17315	9083	1191	5551	3446	6177	5785	10704	5708	1054	14607	8118	1365	9343	10682	8359	10021	10281	11396	2597	17687	10789	6054	1812	6106
Melbourne		13557	16292	16730	14418	11513	16671	8087	16566	16793	2733	13227	15987	12644	11296	13105	8551	8019	6050	15593	711	8159	16264	15762	16381
Mexico City			9842	3728	10740	14834	3362	11701	3603	9213	16272	7669	10260	3038	6585	7422	12071	12945	16623	9603	12972	11319	3262	10172	3033
Milan				6161	2289	5850	6482	9623	6304	643	13689	9242	476	9595	11864	9498	8907	9121	10273	1653	16555	9741	6655	625	6811
Montreal					7077	11692	533	10658	165	5522	18297	8175	6601	4092	8748	8119	10577	11387	14816	5900	16026	10409	502	6453	787
Moscow						6323	7530	7359	7179	2492	12197	11529	2378	9469	14116	11787	6626	6837	8426	1231	14487	7502	7504	1673	7842
Nairobi							11849	10890	11854	6471	8902	8941	5374	15445	11550	9289	10115	9568	7467	6920	12162	11266	12190	5829	12150
New York								11131	541	5851	18702	7729	6907	4140	8218	7656	11078	11897	15349	6336	15990	10870	553	6815	329
Osaka									10598	9660	7654	18695	9735	8664	17614	18754	830	1378	4944	8114	7763	405	10640	9024	11201
Ottawa										5664	18213	8249	6747	3929	8749	8184	10537	11356	14831	6012	15862	10342	351	6587	733
Paris											14268	9146	1108	8975	11628	9376	8990	9288	10743	1546	16959	9738	6015	1038	6180
Perth												13544	13328	14744	12736	13591	7781	7003	3888	13428	3300	7893	18152	13239	18622
Rio de Janeiro													9181	10647	2925	353	18135	18248	15740	10682	13539	18557	8244	9857	7694
Rome														10071	11894	9455	8991	9140	10030	1977	16322	9881	7009	764	7235
San Francisco															9533	10422	9049	9916	13598	8644	11937	8284	3652	9647	3929
Santiago																2590	18361	18865	16399	13067	11369	17234	8585	12490	8036
Sao Paulo																	18337	18568	15998	10904	13378	18526	8159	10116	7603
Seoul																		883	4666	7453	8298	1160	10624	8298	11190
Shanghai																			3783	7788	7847	1782	11460	8499	12025
Singapore																				9646	6293	5317	15015	9717	15555
Stockholm																					15588	8193	6345	1244	6654
Sydney																						7794	15566	15973	15712
Tokyo																							10367	9154	10925
Toronto																								6936	566
Vienna																									7143

MILES

	Madrid	Melbourne	Mexico City	Milan	Montreal	Moscow	Nairobi	New York	Osaka	Ottawa	Paris	Perth	Rio de Janeiro	Rome	San Francisco	Santiago	Sao Paulo	Seoul	Shanghai	Singapore	Stockholm	Sydney	Tokyo	Toronto	Vienna
Melbourne	10759																								
Mexico City	5644	8424																							
Milan	740	10124	6116																						
Montreal	3449	10396	2317	3828																					
Moscow	2141	8959	6674	1422	4398																				
Nairobi	3838	7154	9218	3635	7265	3929																			
New York	3595	10359	2089	4028	331	4679	7363																		
Osaka	6651	5025	7271	5980	6623	4573	6767	6917																	
Ottawa	3547	10294	2239	3917	103	4461	7366	336	6585																
Paris	655	10435	5725	400	3431	1549	4021	3636	6003	3520															
Perth	9077	1698	10111	8506	11370	7579	5532	11621	4756	11317	9090														
Rio de Janeiro	5044	8219	4765	5743	5080	7164	5556	4803	11617	5126	5683	8416													
Rome	848	9934	6375	296	4102	1478	3339	4292	6049	4193	689	8282	5705												
San Francisco	5806	7857	1888	5962	2543	5884	9597	2573	5384	2441	5577	9162	6616	6258											
Santiago	6638	7019	4092	7372	5436	8772	7177	5107	10945	5437	7226	7914	1818	7391	5924										
Sao Paulo	5194	8143	4612	5902	5045	7324	5772	4757	11654	5085	5826	8445	219	5875	6476	1609									
Seoul	6227	5314	7501	5535	6572	4117	6285	6884	516	6548	5586	4835	11269	5587	5623	11409	11394								
Shanghai	6389	4983	8044	5668	7056	4248	5945	7393	856	7057	5771	4352	11339	5680	6162	11722	11538	549							
Singapore	7081	3759	10329	6383	9207	5236	4640	9538	3072	9216	6676	2416	9781	6232	8450	10190	9941	2899	2350						
Stockholm	1614	9689	5967	1027	3666	765	4300	3937	5042	3736	961	8344	6638	1228	5371	8120	6776	4631	4839	5994					
Sydney	10990	442	8061	10287	9958	9002	7557	9936	4824	9856	10538	2051	8413	10142	7418	7065	8313	5156	4876	3910	686				
Tokyo	6704	5070	7034	6053	6468	4662	7001	6755	252	6426	6051	4905	10288	6140	5148	10709	11512	721	1107	3304	5091	4843			
Toronto	3762	10106	2027	4135	312	4663	7575	344	6612	218	3738	11279	5123	4411	2269	5335	5070	6602	7121	9330	3943	9672	6442		
Vienna	1126	9794	6321	388	4010	1040	3622	4235	5607	4093	645	8227	6125	475	5995	7761	6286	5156	5281	6038	773	9925	5688	4310	
Washington	3794	10179	1885	4232	489	4873	7550	204	6960	456	3840	11571	4781	4496	2441	4994	4724	6953	7472	9666	4135	9763	6789	352	4439

KILOMETRES 0 100 200 300 400 500 600 700 800 900 1000 1100 1200 1300 1400 1500 1600

MILES 0 100 200 300 400 500 600 700 800 900 1000

Appendix 7
United States Mileage Chart

Source: Rand McNally, *Interstate Road Atlas*, Chicago: 1989, p. 81

Los Angeles, Calif.	2182	2979	1137	2054	2179	2367	1387	1059	1727	2311	1538	2073	1589		2108	1817	2087	1889	1883	2786	1595	2706	2426	959	1845	715	379	1131	2276	1452	2631	1400
Louisville, Ky.	382	941	1161	292	103	345	819	1120	566	360	928	114	520	2108		367	379	697	685	748	687	668	388	2320	258	1597	2349	2305	301	659	582	710
Mackinaw City, Mich.	935	*916	1291	387	495	439	1281	1341	673	284	1427	460	864	2392	562	880	368	508	1247	906	805	842	556	2128	651	1691	2443	2058	328	1047	775	1061
Madison, Wis.	812	1103	912	144	427	475	968	954	286	406	1137	321	483	2012	432	622	77	272	1006	942	418	878	592	1950	358	1348	2100	1880	372	727	811	676
Memphis, Tenn.	371	1296	1101	530	468	712	452	1040	599	713	561	435	451	1817	367		612	826	390	1100	652	1000	752	2259	285	1535	2125	2290	654	401	867	532
Miami, Fla.	655	1504	2097	1329	1095	1264	1300	2037	1525	1352	1190	1148	1448	2687	1037	997	1416	1723	856	1308	1641	1208	1200	3256	1196	2532	3053	3273	1293	1398	1075	1529
Milwaukee, Wis.	761	1050	987	90	374	422	991	1029	361	353	1142	268	537	2087	379	612		332	994	889	493	825	539	2010	363	1423	2175	1940	319	757	758	734
Minneapolis, Minn.	1068	1368	788	405	692	740	936	920	252	671	1157	586	447	1889	697	826	332		1214	1207	380	1143	857	1678	552	1312	1940	1608	637	695	1076	644
Mobile, Ala.	335	1372	1443	851	706	950	590	1355	954	965	478	716	812	1977	605	363	933	1173	143	1176	1013	1076	982	2587	632	1832	2343	2651	906	713	943	893
Montreal, Que.	1181	318	1773	828	805	561	1705	1815	1146	562	1827	840	1305	2873	906	1273	915	1163	1591	378	1278	449	583	2900	1101	2225	2961	2685	621	1471	579	1502
Nashville, Tenn.	242	1088	1200	446	269	513	660	1156	628	528	769	279	556	2025	175	210	532	826	517	892	744	792	553	2359	299	1636	2333	2376	469	609	659	699
New Orleans, La.	479	1507	1361	912	786	1030	496	1273	978	1045	356	796	806	1883	685	390	994	1214		1311	1007	1211	1070	2505	673	1738	2249	2574	986	647	1078	816
New York, N. Y.	841	222	1746	802	647	473	1552	1771	1119	637	1608	713	1198	2786	748	1100	889	1207	1311		1251	101	368	2885	948	2182	2934	2815	578	1344	233	1394
Norfolk, Va.	540	558	1764	831	604	508	1329	1758	1141	666	1328	700	1162	2694	642	877	918	1236	1019	362	1273	272	384	2914	905	2200	2952	2844	607	1278	188	1352
Oklahoma City, Okla.	839	1641	697	787	840	1029	209	609	544	1013	449	735	349	1349	763	468	861	796	668	1448	455	1368	1088	1841	500	1100	1657	1910	954	105	1293	157
Omaha, Nebr.	986	1412	495	459	693	784	644	537	135	716	865	587	201	1595	687	652	493	380	1007	1251		1183	895	1654	449	931	1683	1638	681	435	1116	298
Orlando, Fla.	435	1294	1876	1109	875	1054	1078	1815	1305	1134	968	928	1226	2465	817	775	1196	1503	634	1098	1421	998	990	3034	976	2310	2831	3053	1075	1176	865	1307
Philadelphia, Pa.	741	296	1678	738	567	413	1452	1691	1051	573	1508	633	1118	2706	668	1000	825	1143	1211	101	1183		288	2821	868	2114	2866	2751	514	1264	143	1314
Phoenix, Ariz.	1793	2604	892	1713	1804	1992	998	792	1409	1957	1149	1698	1214	389	1733	1442	1751	1616	1494	2411	1290	2331	2051	1266	1470	648	763	1437	1901	1077	2256	1025
Pierre, S. Dak.	1361	1726	434	763	1050	1098	943	518	492	1029	1186	944	592	1524	1055	1043	690	394	1394	1565	391	1501	1215	1353	824	823	1575	1283	995	760	1434	578
Pittsburgh, Pa.	687	574	1390	452	295	129	1204	1411	763	287	1313	353	838	2426	388	752	539	857	1070	368	895	288		2535	588	1826	2578	2465	228	984	259	1034
Portland, Me.	1139	115	1986	1042	942	707	1850	2028	1359	775	1906	1001	1486	3074	1043	1398	1129	1447	1609	308	1491	398	663	3125	1236	2422	3174	3055	818	1632	531	1682
Portland, Ore.	2601	3046	1159	2083	2333	2418	2009	1238	1786	2349	2205	2227	1809	959	2320	2259	2010	1678	2505	2885	1654	2821	2535		2060	767	636	174	2315	1913	2754	1739
Raleigh, N. Car.	372	685	1695	784	534	561	1166	1661	1092	683	1160	631	1087	2545	541	728	871	1189	851	489	1214	389	445	2854	831	2131	2853	2797	624	1129	256	1215
Rapid City, S. Dak.	1487	1859	313	896	1177	1231	1050	394	618	1162	1288	1071	708	1363	1182	1159	840	565	1507	1698	507	1634	1348	1204	950	662	1414	1134	1128	873	1567	691
Reno, Nev.	2374	2866	959	1913	2133	2238	1668	1011	1586	2170	1864	2027	1609	469	2120	2003	1946	1711	2164	2705	1454	2637	2349	538	1860	523	229	754	2135	1640	2570	1471
Richmond, Va.	510	535	1674	748	514	425	1266	1668	1058	583	1298	610	1072	2631	552	814	835	1153	989	339	1190	239	301	2831	815	2110	2862	2761	524	1211	105	1262
St. Louis, Mo.	541	1141	901	289	340	529	630	857	333	513	839	246	256	1845	258	285	363	552	673	948	449	868	588	2060		1337	2089	2081	454	396	793	447
Salt Lake City, Utah	1878	2343	436	1390	1610	1715	1242	504	1063	1647	1438	1504	1086	715	1597	1535	1423	1312	1738	2182	931	2114	1826	767	1337		752	836	1612	1172	2047	1003
San Antonio, Texas	983	1988	1027	1187	1160	1404	270	939	954	1392	197	1114	759	1363	1059	692	1261	1206	550	1792	914	1692	1444	2086	900	1319	1737	2155	1333	527	1559	625
San Diego, Calif.	2126	2955	1186	2064	2155	2343	1331	1108	1760	2308	1482	2049	1565	124	2084	1783	2102	1938	1827	2762	1641	2682	2402	1084	1821	764	504	1256	2252	1428	2607	1376
San Francisco, Calif.	2496	3095	1188	2142	2362	2467	1753	1235	1815	2399	1912	2256	1835	379	2349	2125	2175	1940	2249	2934	1683	2866	2578	636	2089	752		808	2364	1760	2799	1695
Seattle, Wash.	2618	2976	1228	2013	2300	2348	2078	1307	1749	2279	2274	2194	1839	1131	2305	2290	1940	1608	2574	2815	1638	2751	2465	174	2081	836	808		2245	1982	2684	1808
Spokane, Wash.	2340	2698	995	1735	2022	2070	1864	1089	1471	2001	2102	1916	1561	1205	2027	2012	1662	1330	2343	2537	1360	2473	2187	348	1803	712	882	278	1967	1726	2406	1544
Springfield, Ill.	592	1099	888	189	299	473	728	865	291	433	879	212	310	1899	275	370	263	480	758	906	412	826	546	2047	97	1324	2076	2039	377	494	751	507
Springfield, Mo.	652	1353	820	499	552	741	419	759	383	725	629	447	192	1636	475	281	573	594	636	1160	371	1080	800	1978	210	1254	1944	2009	666	183	1005	287
Toledo, Ohio	640	739	1176	244	210	116	1084	1218	549	62	1206	219	687	2276	301	654	319	637	986	578	681	514	228	2315	454	1612	2364	2245		850	447	884
Topeka, Kans.	863	1456	595	562	656	844	480	535	255	806	701	550	65	1531	585	508	598	508	863	1263	165	1183	903	1751	322	1028	1770	1802	750	223	1108	139
Tulsa, Okla.	772	1537	765	683	736	925	257	681	443	909	478	631	248	1452	659	401	757	695	647	1344	435	1264	984	1913	396	1172	1760	1982	850		1189	174
Washington, D. C.	608	429	1611	671	481	346	1319	1616	984	506	1375	558	1043	2631	582	867	758	1076	1078	233	1116	143	259	2754	793	2047	2799	2684	447	1189		1239
Wichita, Kans.	903	1587	583	696	787	975	365	509	392	940	608	681	190	1400	710	532	734	644	816	1394	298	1314	1034	1739	447	1003	1695	1808	884	174	1239	

Mileages Copyright ©1987 by Rand McNally—TDM, Inc.

Appendix 8
Political Entities by Size and Population

	Area (Sq. Miles)	Population		Area (Sq. Miles)	Population
Afghanistan	250,775	16,363,000	Faeroe Is. Den.	540	44,000
Africa	11,707,000	434,000,000	Falkland Islands & Dependencies	6,198	1,813
Alabama, U.S.A.	51,705	3,893,978	Fiji	7,055	645,000
Alaska, U.S.A.	591,004	401,851	Finland	130,128	4,812,150
Albania	11,100	2,590,600	Florida, U.S.A.	58,664	9,746,421
Alberta, Canada	255,285	2,237,724	France	210,038	54,334,871
Algeria	919,591	18,666,000	French Guiana	35,135	73,022
American Samoa	77	32,297	French Polynesia	1,544	150,000
Andorra	188	39,940			
Angola	481,351	7,262,000	Gabon	103,346	555,000
Anguilla	35	6,519	Gambia	4,127	601,000
Antarctica	5,500,000		Georgia, U.S.A.	58,910	5,463,087
Antigua and Barbuda	171	75,000	Germany, East (German Democratic Rep.)	41,768	16,732,486
Aruba	75	66,790	Germany, West (Federal Rep.)	95,985	61,546,101
Argentina	1,072,070	28,438,000	Ghana	92,099	11,450,000
Arizona, U.S.A.	114,000	2,718,425	Gibraltar	228	29,648
Arkansas, U.S.A.	53,187	2,286,419	Great Britain and Northern Ireland (U.K.)	93,399	55,638,495
Ascension Island, St. Helena	34	719	Greece	50,944	9,740,417
Asia	17,128,500	2,688,000,000	Greenland	840,000	49,773
Australia	2,966,136	14,576,330	Grenada	133	103,103
Austria	32,375	7,555,338	Guadeloupe & Dep.	687	328,400
			Guam	209	105,979
Bahamas	5,382	209,505	Guatemala	42,042	6,043,559
Bahrain	240	358,857	Guinea	94,925	5,143,284
Bangladesh	55,126	87,052,024	Guinea-Bissau	13,948	810,000
Barbados	166	248,983	Guyana	83,000	793,000
Belgium	11,781	9,848,647			
Belize	8,867	145,353	Haiti	10,694	5,053,792
Benin	43,483	3,338,240	Hawaii, U.S.A.	6,471	964,691
Bermuda	21	67,761	Honduras	43,277	3,955,000
Bhutan	18,147	1,301,000	Hong Kong	403	4,986,560
Bolivia	424,163	5,755,000	Hungary	35,919	10,702,000
Botswana	224,764	936,600			
Brazil	3,284,426	119,098,992	Iceland	39,768	231,000
British Columbia, Canada	366,253	2,744,467	Idaho, U.S.A.	83,564	944,038
British Indian Ocean Terr.	29	2,000	Illinois, U.S.A.	56,345	11,427,414
Brunei	2,226	192,832	India	1,269,339	685,184,692
Bulgaria	42,823	8,890,000	Indiana, U.S.A.	36,185	5,490,260
Burkina Faso	105,869	7,094,000	Indonesia	788,430	147,490,298
Burma	128,769	3,364,000	Iowa, U.S.A.	56,275	2,913,808
Burundi	10,747	4,028,020	Iran	636,293	37,447,000
Byelorussian S.S.R. (White Russian S.S.R.), U.S.S.R.	80,154	9,560,543	Iraq	172,476	12,767,000
			Ireland	21,136	3,443,405
California, U.S.A.	158,706	23,667,837	Ireland, Northern, U.K.	5,452	1,507,065
Cambodia (Kampuchea)	69,898	5,756,141	Isle of Man	227	66,000
Cameroon	183,568	8,503,000	Israel	7,847	3,980,000
Canada	3,851,787	24,343,181	Italy	116,303	56,243,935
Cape Verde	1,557	296,093	Ivory Coast	124,504	7,920,000
Cayman Islands	100	18,000			
Central African Republic	242,000	2,284,000	Jamaica	4,411	2,184,000
Central America	197,480	22,000,000	Japan	145,730	117,060,396
Chad	495,752	4,309,000	Jordan	35,000	2,152,273
Channel Islands	75	129,000			
Chile	292,257	11,275,440	Kalaallit Nunaat (Greenland)	840,000	49,773
China, People's Rep. of	3,691,000	1,008,175,288	Kampuchea (Cambodia)	69,898	5,756,141
China, Republic of (Taiwan)	13,971	18,029,798	Kansas, U.S.A.	82,277	2,364,236
Colombia	439,513	28,776,000	Kentucky, U.S.A.	40,409	3,660,257
Colorado, U.S.A.	104,091	2,889,735	Kenya	224,960	15,327,061
Comoros	719	345,000	Kiribati	291	56,213
Congo	132,046	1,537,000	Korea, North	46,540	18,317,000
Connecticut, U.S.A.	5,018	3,107,576	Korea, South	38,175	37,448,836
Cook Islands	91	17,754	Kuwait	6,532	1,355,827
Costa Rica	19,575	2,271,000			
Cuba	44,206	9,706,369	Laos	91,428	3,811,000
Cyprus	3,473	637,000	Lebanon	4,015	2,688,000
Czechoslovakia	49,373	15,364,000	Lesotho	11,720	1,339,000
			Liberia	43,000	1,873,000
Delaware, U.S.A.	2,044	594,317	Libya	679,358	3,096,000
Denmark	16,629	5,118,000	Liechtenstein	61	25,220
District of Columbia, U.S.A.	69	638,432	Louisiana, U.S.A.	47,752	4,206,098
Djibouti	8,880	386,000	Luxembourg	999	364,606
Dominica	290	74,089			
Dominican Republic	18,704	5,647,977	Macau	6	261,680
			Madagascar	226,657	8,955,000
Ecuador	109,483	8,644,000	Maine, U.S.A.	33,265	1,125,030
Egypt	386,659	43,465,000	Malawi	47,747	6,123,000
El Salvador	8,260	4,748,000	Malaya, Malaysia	50,806	11,138,227
England, U.K.	50,516	46,220,995	Malaysia	128,308	13,435,588
Equatorial Guinea	10,831	244,000	Maldives	115	157,000
Ethiopia	471,776	31,065,000	Mali	464,873	7,160,000
Europe	4,057,000	690,000,000			

Political Entities by Size and Population

	Area (Sq. Miles)	Population		Area (Sq. Miles)	Population
Malta	122	366,000	Saint Vincent & the Grenadines	150	124,000
Manitoba, Canada	250,999	1,026,241	San Marino	23.4	21,000
Martinique	425	328,566	São Tomé e Príncipe	372	86,000
Maryland, U.S.A.	10,460	4,216,941	Sarawak, Malaysia	48,202	1,294,753
Massachusetts, U.S.A.	8,284	5,737,081	Saskatchewan, Canada	251,699	968,313
Mauritania	419,229	1,681,000	Saudi Arabia	829,995	9,319,000
Mauritius	790	971,000	Scotland, U.K.	30,414	5,117,146
Mayotte	144	47,300	Senegal	75,954	5,703,000
Mexico	761,601	67,395,826	Seychelles	145	64,332
Michigan, U.S.A.	58,527	9,262,070	Sierra Leone	27,925	3,571,000
Midway Islands	1.9	468	Singapore	226	2,413,945
Minnesota, U.S.A.	84,402	4,075,970	Solomon Islands	11,500	221,000
Mississippi, U.S.A.	47,689	2,520,631	Somalia	246,200	4,895,000
Missouri, U.S.A.	69,697	4,916,759	South Africa	455,318	23,771,970
Monaco	368 acres	26,000	South America	6,875,000	246,000,000
Mongolia	606,163	1,732,000	South Carolina, U.S.A.	31,113	3,122,814
Montana, U.S.A.	147,046	786,690	South Dakota, U.S.A.	77,116	690,768
Montserrat	40	12,073	South West Africa (Namibia)	317,827	1,009,000
Morocco	172,414	20,646,000	Spain	194,881	37,746,260
Mozambique	303,769	12,130,000	Sri Lanka	25,332	14,850,001
			Sudan	967,494	18,681,000
Namibia (South-West Africa)	317,827	1,009,000	Suriname	55,144	352,041
Nauru	7.7	7,254	Swaziland	6,705	585,000
Nebraska, U.S.A.	77,355	1,569,825	Sweden	173,665	8,328,000
Nepal	54,663	15,020,451	Switzerland	15,943	6,365,960
Netherlands	15,892	14,306,000	Syria	71,498	9,172,000
Netherlands Antilles	390	246,000			
Nevada, U.S.A.	110,561	800,493	Taiwan	13,971	18,029,798
New Brunswick, Canada	28,354	696,403	Tanzania	363,708	17,982,000
New Caledonia &			Tennessee, U.S.A.	42,144	4,591,120
Dependencies	7,335	143,000	Texas, U.S.A.	266,807	14,227,574
Newfoundland, Canada	156,184	567,681	Thailand	198,455	44,278,000
New Hampshire, U.S.A.	9,279	920,610	Togo	21,622	2,702,945
New Jersey, U.S.A.	7,787	7,365,011	Tokelau	3.9	1,625
New Mexico, U.S.A.	121,593	1,303,445	Tonga	270	99,000
New York, U.S.A.	49,108	17,558,072	Trinidad and Tobago	1,980	1,067,108
New Zealand	103,736	3,175,737	Tristan da Cunha, St. Helena	38	323
Nicaragua	45,698	2,732,000	Tunisia	63,378	6,392,000
Niger	489,189	4,994,000	Turkey	300,946	46,312,000
Nigeria	357,000	82,643,000	Turks and Caicos Islands	166	7,436
Niue	100	3,578	Tuvalu	9.78	7,349
North America	9,363,000	376,000,000			
North Carolina, U.S.A.	52,669	5,881,385	Uganda	91,076	12,630,076
North Dakota, U.S.A.	70,702	652,717	Ukrainian S.S.R., U.S.S.R.	233,089	49,754,642
Northern Ireland, U.K.	5,452	1,507,065	Union of Soviet Socialist Reps.	8,649,490	268,800,000
Northwest Territories, Canada	1,304,896	45,741	United Arab Emirates	32,278	1,043,225
Norway	125,053	4,111,000	United Kingdom	94,399	55,638,495
Nova Scotia, Canada	21,425	847,442	United States of America	3,623,420	226,549,448
			Upper Volta (Burkina Faso)	105,869	7,054,000
Ohio, U.S.A.	41,330	10,797,624	Uruguay	72,172	2,947,000
Oklahoma, U.S.A.	69,956	3,025,495	Utah, U.S.A.	84,899	1,461,037
Oman	120,000	919,000			
Ontario, Canada	412,580	8,625,107	Vanuatu	5,700	111,251
Oregon, U.S.A.	97,073	2,633,149	Vatican City	108.7 acres	733
			Venezuela	352,143	14,570,285
Pacific Islands, Terr. of the	533	132,929	Vermont, U.S.A.	9,614	511,456
Pakistan	310,403	83,782,000	Vietnam	128,405	52,741,766
Panama	29,761	1,830,175	Virginia, U.S.A.	40,767	5,346,797
Papua New Guinea	183,540	3,010,727	Virgin Islands, British	59	12,000
Paraguay	157,047	3,026,165	Virgin Islands, U.S.A.	132	96,569
Pennsylvania, U.S.A.	45,308	11,864,751			
Peru	496,222	17,031,221	Wake Island	2.5	302
Philippines	115,707	48,098,460	Wales, U.K.	8,017	2,790,462
Pitcairn Islands	18	54	Wallis and Futuna	106	11,000
Poland	120,725	36,062,309	Washington, U.S.A.	68,139	4,132,204
Portugal	35,549	9,784,200	Western Sahara	102,703	165,000
Prince Edward Island, Canada	2,184	122,506	Western Samoa	1,133	158,130
Puerto Rico	3,515	3,196,520	West Virginia, U.S.A.	24,231	1,950,258
			White Russian S.S.R.		
Qatar	4,247	248,000	(Byelorussian S.S.R.) U.S.S.R.	80,154	9,560,543
Québec, Canada	594,857	6,438,403	Wisconsin, U.S.A.	56,153	4,705,642
			World	54,970,000	4,508,000,000
Réunion	969	515,814	Wyoming, U.S.A.	97,809	469,557
Rhode Island, U.S.A.	1,212	947,154			
Romania	91,699	22,400,000	Yemen, People's Democratic Rep.		
Rwanda	10,169	5,046,000	of	111,101	2,030,000
			Yemen Arab Republic	77,220	6,456,189
Sabah, Malaysia	29,300	1,002,608	Yugoslavia	98,766	22,690,000
Saint Christopher &			Yukon Territory, Canada	207,705	23,153
(St. Kitts)-Nevis	104	44,404			
Saint Helena & Dependencies	162	5,147	Zaire	905,063	26,377,000
Saint Lucia	238	115,783	Zambia	290,586	5,679,808
Saint Pierre & Miquelon	93.5	6,041	Zimbabwe	150,803	7,539,000

Source: Random House, Inc. *The Random House Dictionary of the English Language*, Second Edition, Unabridged, 1987, Atlas, p. 1.

Appendix 9
Metropolitan Areas by Size and Population

Rank in 1989	City	Estimated Mid-Year Populations in Thousands				1989 Area (square miles)	1989 Density (square miles)
		1989	1990	1995	2000		
1.	Tokyo-Yokohama, Japan	26,640	26,952	28,447	29,971	1,089	24,463
2.	Mexico City, Mexico	19,487	20,207	23,913	28,872	522	37,314
3.	Sao Paulo, Brazil	17,376	18,052	21,529	25,354	451	38,528
4.	Seoul, South Korea	15,716	16,268	19,065	21,976	342	45,953
5.	New York, United States	14,618	14,622	14,638	14,648	1,274	11,478
6.	Osaka-Kobe-Kyoto, Japan	13,777	13,826	14,060	14,287	495	27,833
7.	Bombay, India	11,428	11,777	13,532	15,357	95	120,299
8.	Calcutta, India	11,413	11,633	12,885	14,088	209	54,607
9.	Buenos Aires, Argentina	11,360	11,518	12,232	12,911	535	21,233
10.	Rio de Janiero, Brazil	11,153	11,248	12,786	14,169	260	42,894
11.	Moscow, Soviet Union	10,278	10,367	10,769	11,121	379	27,117
12.	Los Angeles, United States	9,974	10,060	10,414	10,714	1,110	6,985
13.	Cairo, Egypt	9,585	9,851	11,155	12,512	104	92,168
14.	Manila, Philippines	9,584	9,880	11,342	12,846	188	50,978
15.	Jakarta, Indonesia	9,275	9,588	11,151	12,804	76	122,033
16.	London, United Kingdom	9,222	9,170	8,897	8,574	874	10,551
17.	Tehran, Iran	8,915	9,354	11,681	14,251	112	79,594
18.	Paris, France	8,693	8,709	8,764	8,803	432	20,123
19.	Delhi, India	8,156	8,475	10,105	11,849	132	59,102
20.	Essen, Germany	7,499	7,474	7,364	7,239	704	10,653
21.	Karachi, Pakistan	7,417	7,711	9,350	11,299	190	39,038
22.	Lagos, Nigeria	7,264	7,602	9,799	12,528	56	129,705
23.	Shanghai, China	6,837	6,873	7,194	7,540	78	87,659
24.	Chicago, United States	6,523	6,526	6,541	6,568	762	8,560
25.	Lima, Peru	6,335	6,578	7,853	9,241	120	54,789
26.	Taipei, Taiwan	6,308	6,513	7,477	8,516	138	45,710
27.	Istanbul, Turkey	6,230	6,461	7,614	8,875	165	37,760
28.	Beijing, China	5,710	5,736	5,865	5,993	151	37,816

Source: *The 1991 Information Please Almanac*, Houghton Mifflin Company, p. 138.

Metropolitan Areas by Size and Population

Rank in 1989	City	Estimated Mid-Year Populations in Thousands				1989 Area (square miles)	1989 Density (square miles)
		1989	1990	1995	2000		
29.	Bangkok, Thailand	5,623	5,791	6,657	7,587	102	55,126
30.	Hong Kong, Hong Kong	5,507	5,656	5,841	5,956	20	28,036
31.	Madras, India	5,582	5,743	6,550	7,384	115	48,541
32.	Bogota, Colombia	5,501	5,710	6,801	7,935	79	69,638
33.	Santiago, Chile	5,154	5,275	5,812	6,294	128	40,269
34.	Tianjin, China	4,767	4,804	5,041	5,298	49	97,291
35.	Milan, Italy	4,727	4,738	4,795	4,839	344	13,741
36.	Nagoya, Japan	4,678	4,736	5,017	5,303	307	15,236
37.	Pusan, South Korea	4,659	4,838	5,748	6,700	54	86,284
38.	Leningrad, Soviet Union	4,653	4,667	4,694	4,738	139	33,474
39.	Bangalore, India	4,410	4,612	5,644	6,764	50	88,191
40.	Madrid, Spain	4,391	4,451	6,772	5,104	66	65,527
41.	Shenyang, China	4,215	4,248	4,457	4,683	39	108,080
42.	Barcelona, Spain	4,101	4,163	4,492	4,834	87	47,138
43.	Lahore, Pakistan	4,101	4,236	4,986	5,864	57	71,949
44.	Manchester, United Kingdom	4,069	4,050	3,949	3,827	357	11,397
45.	Dhaka, Bangladesh	4,016	4,224	5,296	6,492	32	125,511
46.	Philadelphia, United States	4,011	4,007	3,988	3,979	471	8,515
47.	San Francisco, United States	3,924	3,958	4,104	4,214	428	9,167
48.	Baghdad, Iraq	3,813	3,941	4,566	5,239	97	39,304
49.	Ho Chi Minh City, Viet Nam	3,562	3,645	4,046	4,481	31	114,914
50.	Belo Horizonte, Brazil	3,549	3,683	4,373	5,125	79	44,922
51.	Sydney, Australia	3,491	3,515	3,619	3,708	338	10,328
52.	Ahmadabad, India	3,476	3,595	4,200	4,837	32	108,618
53.	Hyderabad, India	3,448	3,563	4,149	4,765	88	39,180
54.	Athens, Greece	3,432	3,468	3,670	3,866	116	29,511
55.	Kinshasa, Zaire	3,403	3,575	4,520	5,646	57	59,704
56.	Miami, United States	3,359	3,421	3,679	3,894	448	7,498
57.	Guangzhou, China	3,313	3,330	3,485	3,652	79	41,942
58.	Surabaya, Indonesia	3,155	3,205	3,428	3,632	43	73,368
59.	Guadalajara, Mexico	3,148	3,262	3,839	4,451	78	40,464
60.	Caracas, Venezuela	3,148	3,188	3,338	3,435	54	58,296
61.	Wuhan, China	3,144	3,169	3,325	3,495	65	48,376
62.	Toronto, Canada	3,080	3,108	3,296	3,295	154	20,001
63.	Greater Berlin, Germany	3,024	3,022	3,018	3,006	274	11,036

Appendix 9

Rank in 1989	City	Estimated Mid-Year Populations in Thousands				1989 Area (square miles)	1989 Density (square miles)
		1989	1990	1995	2000		
64.	Detroit, United States	3,022	2,995	2,685	2,735	468	6,457
65.	Rome, Italy	3,005	3,021	3,079	3,129	69	43,556
66.	Naples, Italy	2,940	2,960	3,051	3,134	62	47,424
67.	Puerto Alegre, Brazil	2,912	3,015	3,541	4,109	231	12,608
68.	Melbourne, Australia	2,896	2,907	2,946	2,968	327	8,856
69.	Montreal, Canada	2,882	2,896	2,996	3,071	164	17,572
70.	Alexandria, Egypt	2,850	2,899	3,144	3,304	35	81,420
71.	Casablanca, Morocco	2,807	2,891	3,327	3,795	35	80,202
72.	Rangoon, Burma	2,760	2,813	3,075	3,332	47	58,724
73.	Monterrey, Mexico	2,730	2,837	3,385	3,974	77	35,448
74.	Kiev, Soviet Union	2,700	2,751	2,983	3,237	62	43,548
75.	Dallas, United States	2,689	2,743	2,972	3,257	419	6,418
76.	Ankara, Turkey	2,687	2,792	3,263	3,777	55	48,852
77.	Singapore, Singapore	2,666	2,695	2,816	2,913	78	34,185
78.	Harbin, China	2,598	2,618	2,747	2,887	30	86,591
79.	Washington, United States	2,529	2,547	2,637	2,207	357	7,083
80.	Boston, United States	2,473	2,475	2,480	2,485	303	8,162
81.	Taegu, South Korea	2,413	2,592	3,201	4,051	n/a	n/a
82.	Lisbon, Portugal	2,366	2,396	2,551	2,717	n/a	n/a
83.	Poona, India	2,351	2,447	2,987	3,647	n/a	n/a
84.	Chengdu, China	2,331	2,349	2,465	2,591	25	93,242
85.	Tashkent, Soviet Union	2,313	2,365	2,640	2,947	n/a	n/a
86.	Budapest, Hungary	2,300	2,301	2,313	2,335	138	16,666
87.	Chongqing, China	2,284	2,339	2,632	2,961	n/a	n/a
88.	Vienna, Austria	2,282	2,313	2,474	2,647	n/a	n/a
89.	Houston, United States	2,258	2,298	2,456	2,651	310	7,284
90.	Birmingham, United Kingdom	2,178	2,170	2,130	2,078	223	9,768
91.	Bucharest, Romania	2,139	2,150	2,214	2,271	52	41,132
92.	Salvador, Brazil	2,123	2,209	2,694	3,266	n/a	n/a
93.	Havana, Cuba	2,087	2,109	2,218	2,333	n/a	n/a
94.	Kanpur, India	2,024	2,076	2,356	2,673	n/a	n/a

Appendix 10
Selected Currencies of the World

Country	Basic Monetary Unit	Intnl. Abbr. or Symbol	Principal Subdivision
Albania	lek	L	100 qintars
Algeria	dinar	DA	100 centimes
Argentina	austral	A	100 centavos
Australia	dollar	A$	100 cents
Austria	schilling	S	100 groschen
Bahamas	dollar	B$	100 cents
Barbados	dollar	BDS$	100 cents
Belgium	franc	BF	100 centimes
Belize	dollar	BZ$	100 cents
Bermuda	dollar	Ber$	100 cents
Bolivia	peso	$b	100 centavos
Brazil	cruzado	Cz	1000 cruzeiros
Britain	pound	£	100 pence
Bulgaria	lev	LV	100 stotinki (sing., *stotinka*)
Canada	dollar	C$	100 cents
Chile	peso	Ch$	100 centavos
China	yuan	RMB¥	100 fen
Colombia	peso	Col$	100 centavos
Congo	CFA franc	CFAF	100 centimes
Costa Rica	colon	₡	100 centimos
Cuba	peso	$	100 centavos
Czechoslovakia	koruna	Kčs	100 halers
Denmark	krone	Dkr	100 öre
Dominican Republic	peso	RD$	100 centavos
East Germany	ostmark	OM	100 pfennigs
Ecuador	sucre	S/	100 centavos
Egypt	pound	£E	100 piasters
El Salvador	colon	₡	100 centavos
Ethiopia	birr	EB	100 cents
Finland	markka	Fmk	100 pennia (sing., *penni*)
France	franc	F	100 centimes
Greece	drachma	Dr	100 lepta (sing., *lepton*)
Guatemala	quetzal	Q	100 centavos
Guyana	dollar	G$	100 cents
Haiti	gourde	G	100 centimes
Honduras	lempira	L	100 centavos
Hong Kong	dollar	HK$	100 cents
Hungary	forint	Ft	100 fillér
Iceland	krona	IKr	100 aurar (sing., *eyrir*)
India	rupee	Re	100 paise (sing., *paisa*)
Indonesia	rupiah	Rp	100 sen
Iran	rial	Rl	100 dinars
Ireland	pound (or punt)	IR£	100 pence
Israel	shekel	IS	100 agorot (sing., *agora*)
Italy	lira	Lit	100 centesimi (sing., *centesimo*)
Ivory Coast	CFA franc	CFAF	100 centimes
Jamaica	dollar	J$	100 cents
Japan	yen	¥	100 sen

Source: Random House, Inc., *The Random House Dictionary of the English Language*, Second Edition, Unabridged, 1987, p. 492.

Appendix 10

Country	Basic Monetary Unit	Intnl. Abbr. or Symbol	Principal Subdivision
Jordan	dinar	JD	1000 fils
Kenya	shilling	KSh	100 cents
Kuwait	dinar	KD	1000 fils
Lebanon	pound	LL	100 piasters
Libya	dinar	LD	1000 dirhams
Luxembourg	franc	LFr	100 centimes
Malawi	kwacha	K	100 tambala
Malaysia	ringgit	M$	100 cents
Mali	franc	MF	100 centimes
Mexico	peso	Mex$	100 centavos
Morocco	dirham	DH	100 centimes
Netherlands	guilder	f	100 cents
New Zealand	dollar	NZ$	100 cents
Nicaragua	cordoba	C$	100 centavos
Nigeria	naira	N	100 kobo
North Korea	won	W	100 chon
Norway	krone	NKr	100 öre
Pakistan	rupee	PRe	100 paise (sing., *paisa*)
Panama	balboa	B	100 centesimos
Paraguay	guarani	₡	100 centimos
Peru	inti	I/	100 cents
Philippines	peso	₱	100 centavos
Poland	zloty	Zl	100 groszy (sing., *grosz*)
Portugal	escudo	Esc	100 centavos
Rumania	leu	L	100 bani (sing., *ban*)
Saudia Arabia	riyal	SRI	100 halalas
Singapore	dollar	S$	100 cents
South Africa	rand	R	100 cents
South Korea	won	W	100 chon
Soviet Union	ruble	Rbl	100 kopecks
Spain	peseta	Pta	100 centimos
Sweden	krona	Skr	100 öre
Switzerland	franc	SwF	100 centimes
Syria	pound	£S	100 piasters
Taiwan	dollar (or yuan)	NT$	100 cents
Thailand	baht	B	100 satangs
Trinidad and Tobago	dollar	TT$	100 cents
Turkey	lira	TL	100 kurus
United Arab Emirates	dirham	Dh	100 fils
United States	dollar	US$	100 cents
Uruguay	peso	NUr$	100 centesimos
Venezuela	bolivar	Bs	100 centimos
Vietnam	dong	D	100 hao
West Germany	Deutsche mark	DM	100 pfennigs
Yugoslavia	dinar	Din	100 paras
Zaire	zaire	Z	100 makuta (sing., *likuta*)
Zimbabwe	dollar	Z$	100 cents

Appendix 11
Greek Alphabet

Α	α	alpha
Β	β	beta
Γ	γ	gamma
Δ	δ	delta
Ε	ε	epsilon
Ζ	ζ	zeta
Η	η	eta
Θ	θ	theta
Ι	ι	iota
Κ	κ	kappa
Λ	λ	lambda
Μ	μ	mu
Ν	ν	nu
Ξ	ξ	xi
Ο	ο	omicron
Π	π	pi
Ρ	ρ	rho
Σ	σ	sigma
Τ	τ	tau
Υ	υ	upsilon
Φ	φ	phi
Χ	χ	chi
Ψ	ψ	psi
Ω	ω	omega

Appendix 12
Mathematical Symbols and Operations

ARITHMETIC AND ALGEBRA

+   1. plus; add. 2. positive; positive value; as: + 64. 3. denoting underestimated approximate accuracy, with some figures omitted at the end; as in: $\pi = 3.14159+$.

−   1. minus; subtract. 2. negative; negative value; as: −64. 3. denoting overestimated approximate accuracy, with some figures omitted at the end; as in: $\pi = 3.141593-$.

±   1. plus or minus; add or subtract; as in: $4 \pm 2 = 6$ or 2. 2. positive or negative; as in: $\sqrt{a^2} = \pm a$. 3. denoting the probable error associated with a figure derived by experiment and observation, approximate calculation, etc.

∓   1. minus or plus. 2. negative or positive, used where ± has appeared previously; as in: $(a \pm b)(a^2 \mp ab + b^2) = (a^3 \pm b^3)$, upper signs or lower signs to be taken consistently throughout.

× ⎰ times; multiplied by; as in: $2 \times 4 =$
· ⎱ $2 \cdot 4$.

Note: Multiplication may also be indicated by writing the algebraic symbols for multiplicand(s) and multiplier(s) close together without any sign; as in: xy; $a(a + b) = a^2 + ab$.

÷ ⎫
/ ⎬ divided by; as in:
− ⎭ $8 \div 2 = 8/2 = \frac{8}{2} = 4$.

: ⎫
/ ⎬ denoting the ratio of (in proportion).
− ⎭

=   equals; is equal to.
:: equals; is equal to (in proportion); as in: $6 : 3 :: 8 : 4$.

≠ ⎫
≠ ⎭ is not equal to.

≡   is identical with.

≢ ⎫
≢ ⎭ is not identical with.

≈   is approximately equal to.

∼   1. is equivalent to. 2. is similar to.
>   is greater than.
≫   is much greater than.
<   is less than.
≪   is much less than.
≯   is not greater than.
≮   is not less than.

≥ ⎫
≧ ⎬ is equal to or greater than.
⩾ ⎭

≤ ⎫
≦ ⎬ is equal to or less than.
⩽ ⎭

∝   varies directly as; is directly proportional to; as in: $x \propto y$.
→   approaches as a limit.

¹, ², ³, etc. (at the right of a symbol, figure, etc.) exponents, indicating the quantity is raised to the first, second, third, etc., power; as in: $(ab)^2 = a^2b^2$; $4^3 = 64$.

√ ⎫ the radical sign, indicating the
√ ⎭ square root of; as in: $\sqrt{81} = 9$.

$\sqrt[3]{\ }$, $\sqrt[4]{\ }$, $\sqrt[5]{\ }$, etc. the radical sign used with indexes, indicating the third, fourth, fifth, etc., root; as in: $\sqrt[3]{125} = 5$.

$^{1/2}$, $^{1/3}$, $^{1/4}$, etc. fractional exponents used to indicate roots, and equal to √, $\sqrt[3]{\ }$, $\sqrt[4]{\ }$, etc.; as in: $9^{1/2} = \sqrt{9} = 3$; $a^{1/6} = a^{1/4} \times a^{1/2} = \sqrt[3]{a^3}$.

⁻¹, ⁻², ⁻³, etc. negative exponents, used to indicate the reciprocal of the same quantity with a positive exponent; as in:
$$9^{-2} = \frac{1}{9^2} = \frac{1}{81}; \ a^{-1/2} = \frac{1}{\sqrt{a}}.$$

( )   parentheses; as in: $2(a + b)$.
[ ]   brackets; as in: $4 + 3[a(a + b)]$.
{ }   braces; as in: $5 + b\{(a + b)[2 - a(a + b)] - 3\}$.
―――   vinculum; as in: $\overline{a + b}$.

Note: Parentheses, brackets, braces, and vinculum are used with quantities consisting of more than one member or term, to group them and show they are to be considered together.

Source: Random House, Inc., *The Random House Dictionary of the English Language*, Second Edition, Unabridged, 1987, pp. 2219–20.

Mathematical Symbols and Operations

! } factorial of; as in:
4! = |4 = 1 · 2 · 3 · 4 = 24.
∞ infinity.
| | absolute value of the quantity within the bars.
% percent; per hundred.
', ", ‴, etc. prime, double prime, triple prime, etc., used to indicate: *a.* constants, as distinguished from the variable denoted by a letter alone. *b.* a variable under different conditions, at different times, etc.
$\log_a x$ logarithm of x to base a.
$\log x$ } common logarithm of x;
$\log_{10} x$ } logarithm of x to base 10.
$\log_e x$ } natural (Naperian) logarithm of x; logarithm of x to
$\ln x$ } base e.
f, F, ϕ, etc. function of; as in: $f(x) = a$ function of x. *Note:* In addition to ϕ, other symbols (esp. letters from the Greek alphabet) may be used to indicate functions, as ψ or γ.
i imaginary unit, $= \sqrt{-1}$.

GEOMETRY

∠
pl. ⩘ } angle; as in: ∠ABC.
⊥   1. a perpendicular *(pl.* ⊥s). 2. is perpendicular to; as in: AB ⊥ CD.
∥   1. a parallel *(pl.* ∥s). 2. is parallel to; as in: AB∥CD.
△
pl. ⩘ } triangle; as in: △ABC.
▭   rectangle; as in :▭ABCD.
◻   square; as in: ◻ABCD.
▱   parallelogram; as in: ▱ABCD.
○
pl. Ⓢ } circle.
≅
≡ } is congruent to; as in:
△ABD ≅ CEF.
∼   is similar to; as in: △ACE ∼ △BDF.
$\stackrel{\text{v}}{=}$ is equiangular.
∴   therefore; hence.
∵   since; because.
$\stackrel{m}{=}$ is measured by.
π   the Greek letter pi, representing the ratio (3.14159+) of the circumference of a circle to its diameter.
—   vinculum, used to indicate: 1. a chord of a circle; as: $\overline{AB}$, the chord between point A and point B. 2. the length of a line segment; as: $\overline{CD}$, the length of the line segment between points C and D.
←   (over a group of letters) indicating

a directed segment; as: $\overleftarrow{EF}$, the directed segment F to E.
⌢   (over a group of letters) indicating an arc of a circle; as: $\overset{\frown}{GH}$, the arc between points G and H.
°   degree(s) of arc; as in: 90°.
′   minute(s) of arc; as in: 90°30′.
″   second(s) of arc; as in: 90°30′15″.

ANALYTIC GEOMETRY

$x \atop y$ } rectangular coordinates of a point on a plane.
$x \atop y \atop z$ } rectangular (Cartesian) coordinates of a point in space.
$r \atop \theta \atop \phi$ } spherical coordinates of a point in space.
$r \atop \theta \atop z$ } cylindrical coordinates of a point in space.
$r \atop \theta$ } polar coordinates of a point in a plane.
C   circumference of a circle.
D   diameter of a circle.
d   perpendicular distance from a point to a line; length of normal.
e   eccentricity.
$l \atop m \atop n$ } direction cosine with X-axis, Y-axis, Z-axis, respectively.
$\cos \alpha \atop \cos \beta \atop \cos \gamma$ } same as l, m, n.
m   slope.
p   semi-latus rectum.
r   radius of a circle.
s   length of arc.
$\alpha \atop \beta \atop \gamma$ } direction angle with X-axis, Y-axis, Z-axis, respectively.
ρ   radius of curvature.

TRIGONOMETRY AND HYPERBOLIC FUNCTIONS

°
′
″ } degree(s), minute(s), second(s), of arc.
θ   angle measured in radians.
sin   sine of angle.
cos   cosine of angle.
tan   tangent of angle.
cot, ctn   cotangent of angle.
sec   secant of angle.
csc   cosecant of angle.
vers   versed sine (versine) of angle; vers $\theta = 1 - \cos \theta$.

covers	coversed sine (coversine) of angle; covers $\theta = 1 - \sin\theta$.
hav	haversine of angle; hav$\theta = \frac{1}{2}(1 - \cos\theta)$.
$\sin^{-1}$, $\cos^{-1}$, etc.	inverse sine of (= angle whose sine is), inverse cosine of (= angle whose cosine is), etc.
arc sin, arc cos, etc.	same as $\sin^{-1}$, $\cos^{-1}$, etc.
sinh, cosh, etc.	hyperbolic sine, hyperbolic cosine, etc.
$\sinh^{-1}$, $\cosh^{-1}$, etc.	inverse hyperbolic sine of (= angle whose hyperbolic sine is), inverse hyperbolic cosine of (= angle whose hyperbolic cosine is), etc.
arc sinh, arc cosh, etc.	same as $\sinh^{-1}$, $\cosh^{-1}$, etc.

CALCULUS

Δ	an increment; as: Δy = an increment of y.
d	differential operator; as: dx = differential of x.
d^2, d^3, d^4, etc.	differential operator of second, third, fourth, etc., order.
D	derivative.
D_x, $\frac{d}{dx}$	derivative with respect to x; as: $D_x y = \frac{dy}{dx}$ = the derivative of y with respect to x.
D_x^2, $\frac{d^2}{dx^2}$	second derivative; as: $D_x^2 y = \frac{d^2 y}{dx^2}$ = the second derivative of y with respect to x.

Note: If $y = f(x)$, the first derivative may also be written as $y' = f'(x)$, the second derivative as $y'' = f''(x)$, etc. In general:

$f^n(x)$, $D_x^n f(x)$, $\frac{d^n f(x)}{dx^n}$	nth derivative of $f(x)$.
∂	partial differential operator.
$\frac{\partial}{\partial x}$	partial derivative with respect to x; as: $\frac{\partial f(x,y)}{\partial x} = D_x f(x,y) = f_x(x,y)$ = the partial derivative of $f(x,y)$ with respect to x.
$\frac{\partial^2 f(x,y)}{\partial y \partial x}$, $D_{yx}^2 f(x,y)$, $f_{yx}(x,y)$	the second partial derivative of $f(x,y)$ with respect to x and then y.
$\dot{x}$, $\ddot{x}$	first and second derivatives with respect to time.
Σ	summation operator.
$\sum_{i=a}^{b}$	summation operator, indicating index (i) and limits (a through b); as in: $\sum_{i=1}^{n} x_i = x_1 + x_2 + \ldots + x_n$.
$\prod_{1}^{n}$	product of n terms.
lim	limit of; as in: $\lim(x) = a$, $y \to b$ indicating the limit of x, as y approaches b, is a.
$\int$	integral sign; as in: $\int f(x)\, dx$, the integral of $f(x)$ with respect to x.
$\int_a^b$	definite integral, giving limits; as in: $\int_a^b f(x)dx$, the definite integral of $f(x)$ between limits a and b.
$F(x)\Big]_a^b$, $F(x)\Big|_a^b$	indicating the difference $F(b) - F(a)$.
$\oint$	integral around a closed path.
γ	Euler's constant (= 0.5772−).

STATISTICS AND PROBABILITY

$P(\ldots)$	probability of $\ldots$; as in: $P(A)$ = probability of A occurring.
$P(A\|B)$	conditional probability; probability of A occurring given that B occurs.
$P(A < X < B)$	probability of X falling between the fixed values A and B.
$\cup$	union; indicating either or both; as in: $P(A \cup B)$ = union of A and B = probability of either A or B or both occurring.
$\cap$	intersection; indicating both; as in: $P(A \cap B)$ = intersection of A and B = probability of both A and B occurring.
$\subset$	is contained in; is a subset of.
$\supset$	implication; contains as a subset.
$-$	(placed over letter) indicates arithmetic mean; as in: $\overline{X}$ = arithmetic mean of X values.
X	1. random variable. 2. an observed value of the random variable X. *Note:* In mathematical and theoretical statistics an observed value is often denoted by a lowercase letter; in applied

Mathematical Symbols and Operations

	statistics the capital letter is generally used.
x	deviation of a value from the sample mean.
$f(X)$	density function; frequency function of the random variable X.
$F(X)$	distribution function; cumulative distribution function.
H:	hypothesis; as in: H: $\mu = 6$, the hypothesis that the population mean is equal to six.
H_0:	null hypothesis; as in: H_0: $p = q$. the hypothesis that no difference exists between p and q.
H_1:	alternative hypothesis; as in: H_1:$p = 2q$.
b	coefficient of regression.
C	contingency coefficient.
df, n	degrees of freedom.
$E(\ldots)$	expected value of. . . .
E	expected frequency in chi-square tests.
O	observed frequency in chi-square tests.
p	proportion of cases in one category of a dichotomous sample; probability of success.
q	$1 - p$; probability of failure.
Q_1	first quartile.
Q_3	third quartile.
$R_{1.23\ldots}$	coefficient of multiple correlation between variable 1 and variables 2, 3, ...
r	product moment correlation coefficient.
r^2	coefficient of determination.
s	1. standard deviation in a sample. 2. standard error of a statistic from sample values (followed by a subscript indicating the statistic).
s^2	sample variance.
V	coefficient of variation.
W	coefficient of concordance.
z	standard score.
α	1. significance level. 2. probability of rejecting a true hypothesis; risk of type I error.
β	1. probability of accepting a hypothesis when an alternative hypothesis is true; risk of type II error. 2. beta coefficient; regression coefficient in standard score form.
η	correlation ratio.
μ	population mean.
μ_1', μ_2', μ_3', etc.	first, second, third, etc., moment of a distribution.
μ_1, μ_2, μ_3, etc.	first, second, third, etc., moment about the mean.
ρ \} R	rank order correlation coefficient.
σ	1. standard deviation. 2. standard error of a statistic (followed by a subscript indicating the statistic).
σ^2	variance.
$\sigma^2(\ldots)$ \} Var $(\ldots)$	variance of. . . .

Appendix 13
Review of Algebraic Operations

1. $a + b = b + a$
2. $a \cdot b = b \cdot a$
3. $a + (b + c) = (a + b) + c$
4. $a \cdot (b \cdot c) = (a \cdot b) \cdot c$
5. $a \cdot (b + c) = a \cdot b + a \cdot c$
6. $a - (b + c) = a - b - c$
7. $\dfrac{a}{b} \cdot \dfrac{c}{d} = \dfrac{ac}{bd}$
8. $\dfrac{a/b}{c/d} = \dfrac{a}{b} \cdot \dfrac{d}{c} = \dfrac{ad}{bc}$
9. $\dfrac{a+b}{c} = \dfrac{a}{c} + \dfrac{b}{c}$
10. $a^m \cdot a^n = a^{m+n}$
11. $\dfrac{a^m}{a^n} = a^{m-n}$
12. $a^{-n} = \dfrac{1}{a^n}$
13. $(a^m)^n = a^{m \cdot n}$
14. $(a \cdot b)^n = a^n \cdot b^n$
15. $\left(\dfrac{a}{b}\right)^n = \dfrac{a^n}{b^n}$
16. $a^0 = 1$
17. $a^{1/n} = \sqrt[n]{a}$
18. $a^{m/n} = (\sqrt[n]{a})^m$
19. $(a + b)^2 = a^2 + 2ab + b^2$
20. $(a - b)^2 = a^2 - 2ab + b^2$
21. $(a + b)^3 = a^3 + 3a^2b + 3ab^2 + b^3$
22. $(a - b)^3 = a^3 - 3a^2b + 3ab^2 - b^3$
23. $(a + b)(a - b) = a^2 - b^2$
24. If $a^x = N$ then $x = \log_a N$
25. $\log_a (M \cdot N) = \log_a M + \log_a N$
26. $\log_a \dfrac{M}{N} = \log_a M - \log_a N$
27. $\log_a N^p = p \cdot \log_a N$
28. $\sum_{i=1}^{N} a_i = a_1 + a_2 + a_3 + \cdots + a_N$
29. $\sum_{i=1}^{N} a_i^2 = a_1^2 + a_2^2 + a_3^2 + \cdots + a_N^2$
30. $\sum_{i=1}^{N} a_i b_i = a_1 b_1 + a_2 b_2 + a_3 b_3 + \cdots + a_N b_N$
31. $\sum_{i=1}^{N} (a_i + b_i) = (a_1 + b_1) + (a_2 + b_2) + (a_3 + b_3) + \cdots + (a_N + b_N)$
$= \sum_{i=1}^{N} a_i + \sum_{i=1}^{N} b_i$
32. $\sum_{i=1}^{N} (a_i + b_i)^2 = \sum_{i=1}^{N} (a_i^2 + 2a_i b_i + b_i^2)$
$= \sum_{i=1}^{N} a_i^2 + \sum_{i=1}^{N} 2a_i b_i + \sum_{i=1}^{N} b_i^2$
33. $\sum_{i=1}^{N} k a_i = k \sum_{i=1}^{N} a_i$ where k is a constant
34. $\sum_{i=1}^{N} \sum_{j=1}^{M} a_{ij} = \sum_{j=1}^{M} a_{1j} + \sum_{j=1}^{M} a_{2j} + \cdots + \sum_{j=1}^{M} a_{Nj}$

Source: Donald A. Krueckeberg and Arthur Silvers, *Urban Planning Analysis: Methods and Models.* New York : John Wiley and Sons, Inc., 1974, p. 441.

Appendix 14
Present Value Tables

Discount Rate																
Year	1%	2%	3%	4%	5%	6%	7%	8%	9%	10%	12%	15%	20%	25%	30%	40%	50%
1	0.9901	0.9804	0.9709	0.9615	0.9524	0.9434	0.9346	0.9259	0.9174	0.9091	0.8929	0.8696	0.8333	0.8000	0.7692	0.7143	0.6667
2	0.9803	0.9707	0.9613	0.9520	0.9430	0.9341	0.9253	0.9168	0.9083	0.9001	0.8840	0.8610	0.8251	0.7921	0.7616	0.7072	0.6601
3	0.9706	0.9611	0.9517	0.9426	0.9336	0.9248	0.9162	0.9077	0.8994	0.8912	0.8753	0.8524	0.8169	0.7842	0.7541	0.7002	0.6535
4	0.9610	0.9516	0.9423	0.9333	0.9244	0.9157	0.9071	0.8987	0.8904	0.8824	0.8666	0.8440	0.8088	0.7765	0.7466	0.6933	0.6471
5	0.9515	0.9421	0.9330	0.9240	0.9152	0.9066	0.8981	0.8898	0.8816	0.8735	0.8580	0.8356	0.8008	0.7688	0.7392	0.6864	0.6407
6	0.9420	0.9328	0.9238	0.9149	0.9062	0.8976	0.8892	0.8810	0.8729	0.8650	0.8495	0.8274	0.7929	0.7612	0.7319	0.6796	0.6343
7	0.9327	0.9236	0.9146	0.9058	0.8972	0.8887	0.8804	0.8723	0.8643	0.8564	0.8411	0.8192	0.7850	0.7536	0.7247	0.6729	0.6280
8	0.9235	0.9144	0.9056	0.8968	0.8883	0.8799	0.8717	0.8636	0.8557	0.8479	0.8328	0.8111	0.7773	0.7462	0.7175	0.6662	0.6218
9	0.9143	0.9054	0.8966	0.8880	0.8795	0.8712	0.8631	0.8551	0.8472	0.8395	0.8245	0.8030	0.7696	0.7388	0.7104	0.6596	0.6157
10	0.9053	0.8964	0.8877	0.8792	0.8708	0.8626	0.8545	0.8466	0.8388	0.8312	0.8164	0.7951	0.7619	0.7315	0.7033	0.6531	0.6096
11	0.8963	0.8875	0.8789	0.8705	0.8622	0.8540	0.8461	0.8382	0.8305	0.8230	0.8083	0.7872	0.7544	0.7242	0.6964	0.6466	0.6035
12	0.8874	0.8787	0.8702	0.8618	0.8536	0.8456	0.8377	0.8299	0.8223	0.8148	0.8003	0.7794	0.7469	0.7171	0.6895	0.6402	0.5975
13	0.8787	0.8700	0.8616	0.8533	0.8452	0.8372	0.8294	0.8217	0.8142	0.8068	0.7924	0.7717	0.7395	0.7100	0.6827	0.6339	0.5916
14	0.8700	0.8614	0.8531	0.8449	0.8368	0.8289	0.8212	0.8136	0.8061	0.7988	0.7845	0.7641	0.7322	0.7029	0.6759	0.6278	0.5858
15	0.8613	0.8529	0.8446	0.8365	0.8285	0.8207	0.8130	0.8055	0.7981	0.7909	0.7768	0.7565	0.7250	0.6960	0.6692	0.6214	0.5800
16	0.8528	0.8445	0.8363	0.8282	0.8203	0.8126	0.8050	0.7975	0.7902	0.7830	0.7691	0.7490	0.7178	0.6891	0.6626	0.6152	0.5742
17	0.8444	0.8361	0.8280	0.8200	0.8122	0.8045	0.7970	0.7896	0.7824	0.7753	0.7614	0.7416	0.7107	0.6823	0.6560	0.6092	0.5685
18	0.8360	0.8278	0.8198	0.8119	0.8042	0.7966	0.7891	0.7818	0.7747	0.7676	0.7539	0.7342	0.7036	0.6755	0.6495	0.6031	0.5629
19	0.8277	0.8196	0.8117	0.8039	0.7962	0.7887	0.7813	0.7741	0.7670	0.7600	0.7464	0.7270	0.6967	0.6688	0.6431	0.5972	0.5573
20	0.8195	0.8115	0.8036	0.7959	0.7883	0.7809	0.7736	0.7664	0.7594	0.7525	0.7391	0.7198	0.6898	0.6622	0.6367	0.5912	0.5518
21	0.8114	0.8035	0.7957	0.7880	0.7805	0.7732	0.7659	0.7588	0.7519	0.7450	0.7317	0.7126	0.6830	0.6556	0.6304	0.5854	0.5464
22	0.8034	0.7955	0.7878	0.7802	0.7728	0.7655	0.7583	0.7513	0.7444	0.7377	0.7245	0.7056	0.6762	0.6491	0.6242	0.5796	0.5410
23	0.7954	0.7876	0.7800	0.7725	0.7651	0.7579	0.7508	0.7439	0.7371	0.7304	0.7173	0.6986	0.6695	0.6427	0.6180	0.5739	0.5356
24	0.7876	0.7798	0.7723	0.7648	0.7576	0.7504	0.7434	0.7365	0.7298	0.7231	0.7102	0.6917	0.6629	0.6364	0.6119	0.5682	0.5303
25	0.7798	0.7721	0.7646	0.7573	0.7501	0.7430	0.7360	0.7292	0.7225	0.7160	0.7032	0.6848	0.6563	0.6301	0.6058	0.5625	0.5250
30	0.7419	0.7346	0.7275	0.7205	0.7137	0.7069	0.7003	0.6938	0.6875	0.6812	0.6691	0.6516	0.6245	0.5995	0.5764	0.5352	0.4996
40	0.6717	0.6651	0.6586	0.6523	0.6461	0.6400	0.6340	0.6281	0.6224	0.6167	0.6057	0.5899	0.5653	0.5427	0.5218	0.4845	0.4522
50	0.6080	0.6021	0.5962	0.5905	0.5849	0.5794	0.5739	0.5686	0.5634	0.5583	0.5483	0.5340	0.5118	0.4913	0.4724	0.4387	0.4094

The numbers in this table answer the question: "What is the value today of $1 given to me in the year n, if the discount rate employed is r%?"

Source: Carl V. Patton and David S. Sawicki, *Basic Methods of Policy Analysis and Planning*, Englewood Cliffs, NJ: Prentice-Hall, Inc., 1986, pp. 258–59.

Appendix 14

Discount Rate

Year	1%	2%	3%	4%	5%	6%	7%	8%	9%	10%	12%	15%	20%	25%	30%	40%	50%
1	0.9901	0.9804	0.9709	0.9615	0.9524	0.9434	0.9346	0.9259	0.9174	0.9091	0.8929	0.8696	0.8333	0.8000	0.7692	0.7143	0.6667
2	1.9704	1.9416	1.9135	1.8861	1.8594	1.8334	1.8080	1.7833	1.7591	1.7355	1.6901	1.6257	1.5278	1.4400	1.3609	1.2245	1.1111
3	2.9410	2.8839	2.8286	2.7751	2.7232	2.6730	2.6243	2.5771	2.5313	2.4869	2.4018	2.2832	2.1065	1.9520	1.8161	1.5889	1.4074
4	3.9020	3.8077	3.7171	3.6299	3.5460	3.4651	3.3872	3.3121	3.2397	3.1699	3.0373	2.8550	2.5887	2.3616	2.1662	1.8492	1.6049
5	4.8534	4.7135	4.5797	4.4518	4.3295	4.2124	4.1002	3.9927	3.8897	3.7908	3.6048	3.3522	2.9906	2.6893	2.4356	2.0352	1.7366
6	5.7955	5.6014	5.4172	5.2421	5.0757	4.9173	4.7665	4.6229	4.4859	4.3553	4.1114	3.7845	3.3255	2.9514	2.6427	2.1680	1.8244
7	6.7282	6.4720	6.2303	6.0021	5.7864	5.5824	5.3893	5.2064	5.0330	4.8684	4.5638	4.1604	3.6046	3.1611	2.8021	2.2628	1.8829
8	7.6517	7.3255	7.0197	6.7327	6.4632	6.2098	5.9713	5.7466	5.5348	5.3349	4.9676	4.4873	3.8372	3.3289	2.9247	2.3306	1.9220
9	8.5660	8.1622	7.7861	7.4353	7.1078	6.8017	6.5152	6.2469	5.9952	5.7590	5.3282	4.7716	4.0310	3.4631	3.0190	2.3790	1.9480
10	9.4713	8.9826	8.5302	8.1109	7.7217	7.3601	7.0236	6.7101	6.4177	6.1446	5.6502	5.0188	4.1925	3.5705	3.0915	2.4136	1.9653
11	10.3676	9.7868	9.2526	8.7605	8.3064	7.8869	7.4987	7.1390	6.8052	6.4951	5.9377	5.2337	4.3271	3.6564	3.1473	2.4383	1.9769
12	11.2551	10.5753	9.9540	9.3851	8.8633	8.3838	7.9427	7.5361	7.1607	6.8137	6.1944	5.4206	4.4392	3.7251	3.1903	2.4559	1.9846
13	12.1337	11.3484	10.6350	9.9856	9.3936	8.8527	8.3577	7.9038	7.4869	7.1034	6.4235	5.5831	4.5327	3.7801	3.2233	2.4685	1.9897
14	13.0037	12.1062	11.2961	10.5631	9.8986	9.2950	8.7455	8.2442	7.7862	7.3667	6.6282	5.7245	4.6106	3.8241	3.2487	2.4775	1.9931
15	13.8651	12.8493	11.9379	11.1184	10.3797	9.7122	9.1079	8.5595	8.0607	7.6061	6.8109	5.8474	4.6755	3.8593	3.2682	2.4839	1.9954
16	14.7179	13.5777	12.5611	11.6523	10.8378	10.1059	9.4466	8.8514	8.3126	7.8237	6.9740	5.9542	4.7296	3.8874	3.2832	2.4885	1.9970
17	15.5623	14.2919	13.1661	12.1657	11.2741	10.4773	9.7632	9.1216	8.5436	8.0216	7.1196	6.0472	4.7746	3.9099	3.2948	2.4918	1.9980
18	16.3983	14.9920	13.7535	12.6593	11.6896	10.8276	10.0591	9.3719	8.7556	8.2014	7.2497	6.1280	4.8122	3.9279	3.3037	2.4941	1.9986
19	17.2260	15.6785	14.3238	13.1339	12.0853	11.1581	10.3356	9.6036	8.9501	8.3649	7.3658	6.1982	4.8435	3.9424	3.3105	2.4958	1.9991
20	18.0456	16.3514	14.8775	13.5903	12.4622	11.4699	10.5940	9.8181	9.1285	8.5136	7.4694	6.2593	4.8696	3.9539	3.3158	2.4970	1.9994
21	18.8570	17.0112	15.4150	14.0292	12.8212	11.7641	10.8355	10.0168	9.2922	8.6487	7.5620	6.3125	4.8913	3.9631	3.3198	2.4979	1.9996
22	19.6604	17.6580	15.9369	14.4511	13.1630	12.0416	11.0612	10.2007	9.4424	8.7715	7.6446	6.3587	4.9094	3.9705	3.3230	2.4985	1.9997
23	20.4558	18.2922	16.4436	14.8568	13.4886	12.3034	11.2722	10.3711	9.5802	8.8832	7.7184	6.3988	4.9245	3.9764	3.3254	2.4989	1.9998
24	21.2434	18.9139	16.9355	15.2470	13.7986	12.5504	11.4693	10.5288	9.7066	8.9847	7.7843	6.4338	4.9371	3.9811	3.3272	2.4992	1.9999
25	22.0232	19.5235	17.4131	15.6221	14.0939	12.7834	11.6536	10.6748	9.8226	9.0770	7.8431	6.4641	4.9476	3.9849	3.3286	2.4994	1.9999
30	25.8077	22.3965	19.6004	17.2920	15.3725	13.7648	12.4090	11.2578	10.2737	9.4269	8.0552	6.5660	4.9789	3.9950	3.3321	2.4999	2.0000
40	32.8347	27.3555	23.1148	19.7928	17.1591	15.0463	13.3317	11.9246	10.7574	9.7791	8.2438	6.6418	4.9966	3.9995	3.3332	2.5000	2.0000
50	39.1961	31.4236	25.7298	21.4822	18.2559	15.7619	13.8007	12.2335	10.9617	9.9148	8.3045	6.6605	4.9995	3.9999	3.3333	2.5000	2.0000

The numbers in this table answer the question: "What is the value today of $1 given to me each year for n years, if the discount rate employed is r%?"

Appendix 15
Distance Conversions

Customary Units		Metric Units
1 parsec	=	30.84×10^{15} m
1 light year	=	9 464 Tm
1 league	=	4.828 km
1 nautical mile	=	1.852 km
1 mile	=	1.609 km
1 chain	=	20.12 m
1 rod	=	5.029 m
1 fathom	=	1.83 m
1 yard (yd)	=	0.914 m
1 foot (ft)	=	0.3048 m
1 hand	=	101.6 mm
1 inch (in)	=	25.4 mm
1 point (printer's)	=	0.353 mm
1 micron	=	1 μm
1 angstrom (Å)	=	0.1 nm

Useful approximations:

1 km	=	5/8 mile
10 cm	=	4 in
1 mm	=	40 thousandths
10 ft	=	3 m
2.5 cm	=	1 in
25 mm	=	1 in
30 cm	=	1 ft

[Conversion scale chart showing MILLIMETRES, INCHES, METRES, YARDS, KILOMETRES, MILES]

Source: John L. Frirer, *SI Metric Handbook*, New York: Charles Scribner's Sons, 1977, Figure 4-13.

Appendix 16
Area Comparisons and Conversions

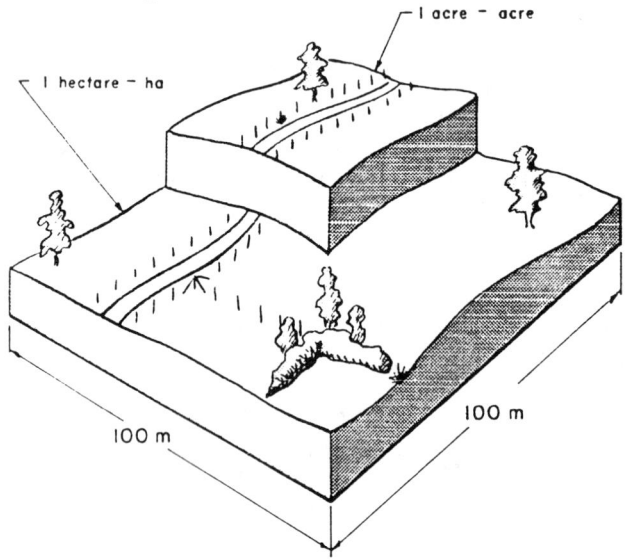

Farms and other plots of land are shown in hectares.

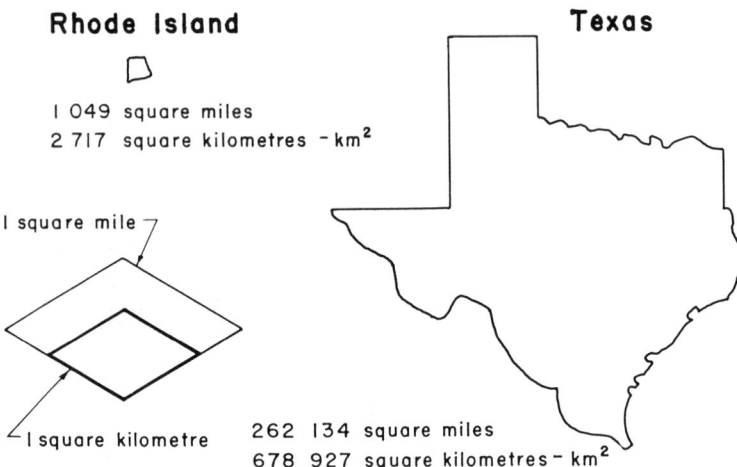

Very large areas, such as the sizes of states, are shown in square kilometres.

Sources: John L. Frirer, *SI Metric Handbook*, New York: Charles Scribner's Sons, 1977, Figures 5-6, 5-7, 5-8, 5-10, 5-11a, b, and American Appraisal Company, *Boeckh Building Valuation Manual, Volume II, Commercial*, Milwaukee, WI: 1967, pp. 380–381.

Area Comparisons and Conversions

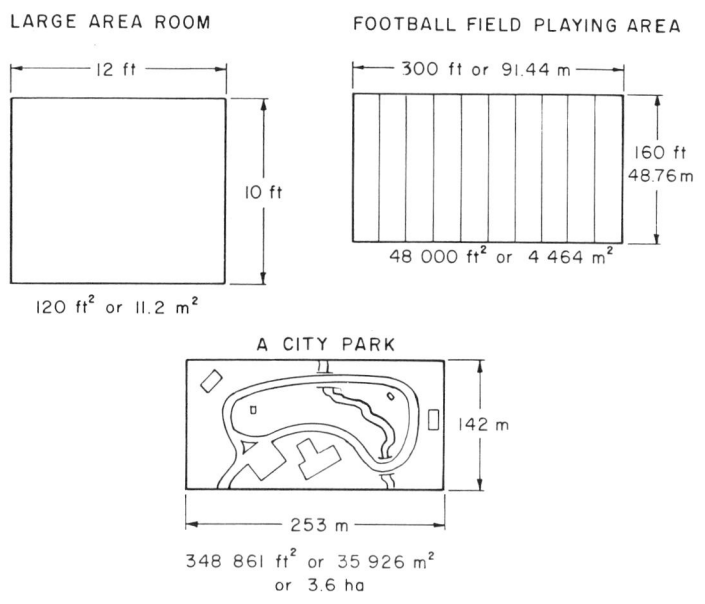

Larger areas are shown in square metres.

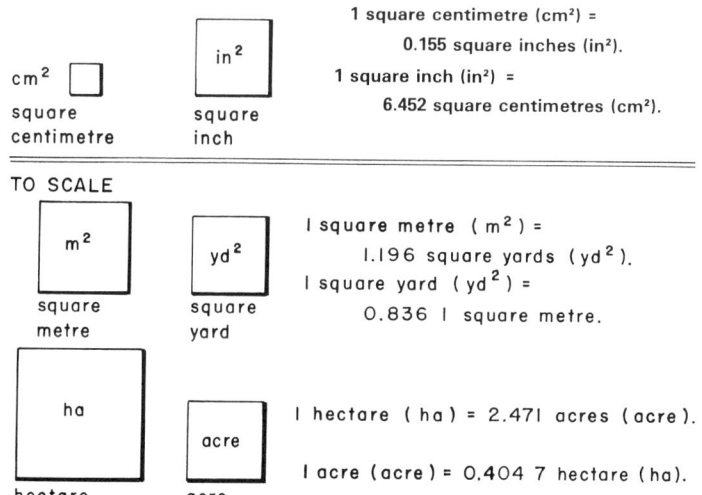

A comparison between common square measures in metric and in customary.

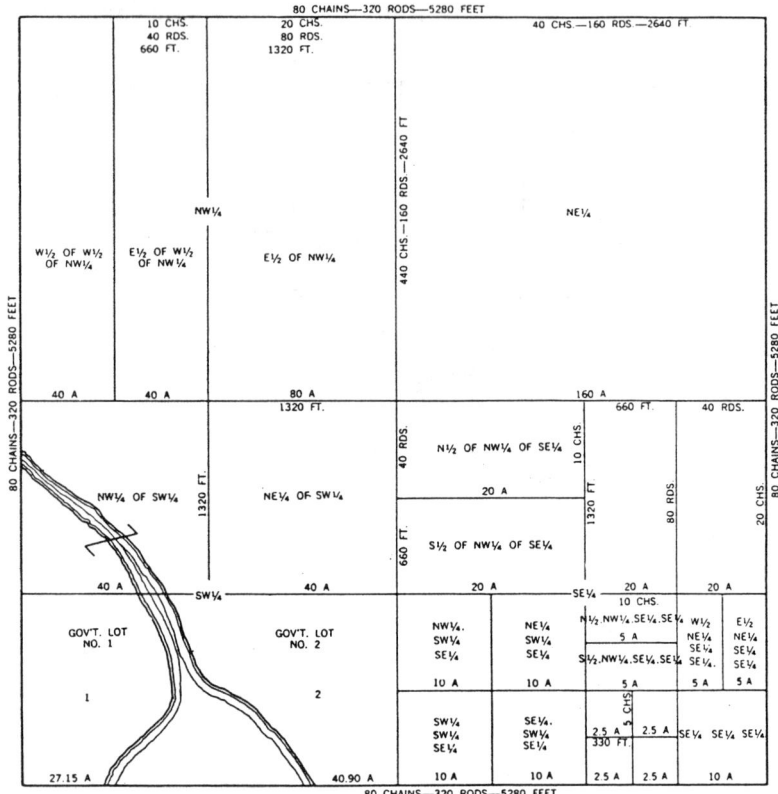

The above figure illustrates a section of land 640 acres divided first into quarters of 160 acres each, shown as NE1/4, NW1/4, SW1/4, SE1/4, and then into various other divisions.

Area Comparisons and Conversions

LINEAL MEASURE

1 mile	=	5,280	feet
	=	1,760	yards
	=	320	rods
	=	80	chains
1 chain	=	66	feet
	=	100	links
	=	4	rods
1 rod	=	25	links
	=	16.5	feet
	=	1	perch
	=	1	pole
1 link	=	7.92	inches

Millimeter = 0.001 Meter
Centimeter = 0.01 Meter
Decimeter = 0.1 Meter
Meter = 39.3685 Inches
Kilometer = 1000 Meters

AREA MEASURE

1 township	=	36	sections
1 full section	=	640	acres
1 sq. mile	=	640	acres
	=	1	full section
1 acre	=	43,560	sq. feet
	=	4,840	sq. yards
	=	160	sq. rods
	=	10	sq. chains
1 sq. chain	=	10,000	sq. links
1 sq. rod	=	30.25	sq. yards
1 sq. yard	=	9	sq. feet
1 sq. foot	=	144	sq. inches

Square Centimeter = 0.0001 Square Meter
Square Decimeter = 0.01 Square Meter
Are = 100 Square Meters
Hectare = 10,000 Square Meters
= 2.471 Acres
Square Kilometer = 247.1 Acres
= 0.386 Square Mile

ARPENT

The Arpent is a unit of measure common to parts of Canada, mainly Quebec, where land was originally granted under seigniorial tenure. Surveys currently made in these areas now use the English units, but the Arpent may be encountered.

This unit is also in use in parts of the State of Louisiana.

The basis of the Arpent is the "Old French Foot" having the following equivalents:

French Foot = 12.789 English Inches
= 1.06575 English Feet
English Foot = 12 English (U.S.) Inches
= 0.938306 French Feet
Square
French Foot = 1.135823 Square English Feet
Lineal Arpent = 180 French Feet
= 191.835 English Feet
= 10 Old French Perches
Square Arpent = 36800.667 Square English Feet
= 4088.89 Square English Yards
= 32400 Square French Feet
= 0.845 U.S. Acre

The Old French Perch is equivalent to 18 French Feet or 19.1835 English Feet.

VARA

The Vara is a unit of measurement originally used by the Spanish and is still in common use throughout Central and South America.

The exact length of the Vara range varies from 32.9931 to 34.1208 inches with each country using a variation within this range.

Within the United States, two areas still make use of this measurement unit.

California

Vara is equal to 33 inches.

Many lots in San Francisco and other areas were laid out on the basis of multiples of 50 Varas (137'6").

Texas

Vara is equal to 33.33333 inches.

Early deeds used "Leagues" and "Labors" having the following values:

League = 4428.4 Acres
= 5000 Varas Square
= 25,000,000 Square Varas

Labor = 177.1 Acres
= 1000 Varas Square
= 1,000,000 Square Varas

Conversion of these two variations of the Vara in standard United States units can be found in the conversion tables.

Appendix 16

Area

	hectare	square metre	square millimetre	square mile	acre	square yard	square foot	square inch
1 hectare	1.0	1 000.0	---	0.004*	2.471	---	---	---
1 square metre	---	1.0	---	---	---	1.196	10.764	1 550.003
1 square mile	258.999	---	---	1.0	640.0	---	---	---
1 acre	0.405	4 046.856	---	---	1.0	4 840.0	---	---
1 square yard	---	0.836	---	---	---	1.0	9.0	1 296.0
1 square foot	---	0.093*	92 903.04	---	---	0.111	1.0	144.0
1 square inch	---	---	645.16	---	---	---	0.007*	1.0

*Round off. Use with caution

Units of Area

To Convert from square centimetres

To	Multiply by
square inches	0.155 000 3
square feet	0.0001 076 39
square yards	0.000 119 599
square metres	0.000 1

To Convert from hectares

To	Multiply by
square feet	107 639.1
square yards	11 959.90
acres	2.471 054
square miles	0.003 861 02
square metres	10 000

To Convert from square feet

To	Multiply by
square inches	144
square yards	0.111 111 1
acres	0.000 022 957
square centimetres	929.030 4
square metres	0.092 903 04

To Convert from square inches

To	Multiply by
square feet	0.006 944 44
square yards	0.000 771 605
square centimetres	6.451 6
square metres	0.000 645 16

To Convert from square metres

To	Multiply by
square inches	1 550.003
square feet	10.763 91
square yards	1.195 990
acres	0.000 247 105
square centimetres	10 000
hectares	0.000 1

To Convert from square yards

To	Multiply by
square inches	1 296
square feet	9
acres	0.000 206 611 6
square miles	0.000 000 322 830 6
square centimetres	8 361.273 6
square metres	0.836 127 36
hectares	0.000 083 612 736

To Convert from square miles

To	Multiply by
square feet	27 878 400
square yards	3 097 600
acres	640
square metres	2 589 988.110 336
hectares	258.998 811 033 6

To Convert from acres

To	Multiply by
square feet	43 560
square yards	4 840
square miles	0.001 562 5
square metres	4 046.856 422 4
hectares	0.404 685 642 24

Appendix 17
Volume Conversions

To	To Convert from cubic feet	Multiply by
liquid ounces	957.506 5	
gills	239.376 6	
liquid pints	59.844 16	
liquid quarts	29.922 08	
gallons	7.480 519	
cubic inches	1 728	
litres	28.316 846 592	
cubic metres	0.028 316 846 592	
cubic yards	0.037 037 04	

To	To Convert from liquid (fluid) quarts	Multiply by
liquid ounces	32	
liquid pints	2	
gallons	0.25	
cubic inches	57.75	
cubic feet	0.033 420 14	
millilitres	946.352 946	
litres	0.946 352 946	

To	To Convert from cubic inches	Multiply by
liquid ounces	0.554 112 6	
gills	0.138 528 1	
liquid pints	0.034 632 03	
liquid quarts	0.017 316 02	
gallons	0.004 329 0	
cubic feet	0.000 578 7	
millilitres	16.387 064	
litres	0.016 387 064	
cubic metres	0.000 016 387 064	
cubic yards	0.000 021 43	

To	To Convert from gallons	Multiply by
liquid ounces	128	
liquid pints	8	
liquid quarts	4	
cubic inches	231	
cubic feet	0.133 680 6	
millilitres	3 785.411 784	
litres	3.785 411 784	
cubic metres	0.003 785 411 784	
cubic yards	0.004 951 13	

Source: John L. Frirer, *SI Metric Handbook*, New York: Charles Scribner's Sons, 1977, Figures 6-12b, c.

Units of Capacity, or Volume, Dry Measure

To Convert from litres	
To	Multiply by
dry pints	1.816 166
dry quarts	0.908 082 98
pecks	0.113 510 4
bushels	0.028 377 59

To Convert from cubic metres	
To	Multiply by
pecks	113.510 4
bushels	28.377 59

To Convert from dry pints	
To	Multiply by
dry quarts	0.5
pecks	0.062 5
bushels	0.015 625
cubic inches	33.600 312 5
cubic feet	0.019 444 63
litres	0.550 610 47

To Convert from dry quarts	
To	Multiply by
dry pints	2
pecks	0.125
bushels	0.031 25
cubic inches	67.200 625
cubic feet	0.038 889 25
litres	1.101 221

To Convert from bushels	
To	Multiply by
dry pints	64
dry quarts	32
pecks	4
cubic inches	2 150.42
cubic feet	1.244 456
litres	35.239 07
cubic metres	0.035 239 07
cubic yards	0.046 090 96

To Convert from cubic feet	
To	Multiply by
dry pints	51.428 09
dry quarts	25.714 05
pecks	3.214 256
bushels	0.803 563 95

To Convert from pecks	
To	Multiply by
dry pints	16
dry quarts	8
bushels	0.25
cubic inches	537.605
cubic feet	0.311 114
litres	8.809 767 5
cubic metres	0.008 809 77
cubic yards	0.011 522 74

To Convert from cubic inches	
To	Multiply by
dry pints	0.029 761 6
dry quarts	0.014 880 8
pecks	0.001 860 10
bushels	0.000 465 025

To Convert from cubic yards	
To	Multiply by
pecks	86.784 91
bushels	21.696 227

Appendix 18
Weight Comparisons and Conversions

```
━━━WEIGHT (MASS) IN GRAMS━━━
```

DIME
2.5 g

NICKEL
5 g

QUARTER
6.25 g

HALF DOLLAR
12.5 g

RADIO BATTERY
9 VOLTS
40 g

GOLF BALL
50 g

FOOD
TUNA
250 g

Source: John L. Frirer, *SI Metric Handbook*, New York: Charles Scribner's Sons, 1977, Figures 7-5, 7-9c.

Units of Mass or Weight

To Convert from grams	
To	Multiply by
ounces	0.035 273 96
pounds	0.002 204 62
milligrams	1 000
kilograms	0.001

To Convert from metric tons (tonne)	
To	Multiply by
pounds	2 204.623
short tons	1.102 311 3
long tons	0.964 206 5
kilograms	1 000

To Convert from pounds	
To	Multiply by
ounces	16
kilograms	0.453 592 37
short tons	0.000 5
long tons	0.000 446 428 6
metric tons (tonne)	0.000 453 592 37

To Convert from short tons	
To	Multiply by
pounds	2 000
long tons	0.892 857 1
kilograms	907.184 74
metric tons (tonne)	0.907 184 74

To Convert from kilograms	
To	Multiply by
ounces	35.273 96
pounds	2.204 623
grams	1 000
short tons	0.001 102 31
long tons	0.000 984 2
metric tons (tonne)	0.001

To Convert from ounces	
To	Multiply by
pounds	0.062 5
grams	23.349 523 125
kilograms	0.028 349 523 125

To Convert from long tons	
To	Multiply by
ounces	35 840
pounds	2 240
short tons	1.12
kilograms	1 016.046 908 8
metric tons (tonne)	1.016 046 908 8

Appendix 19
Temperature Conversions

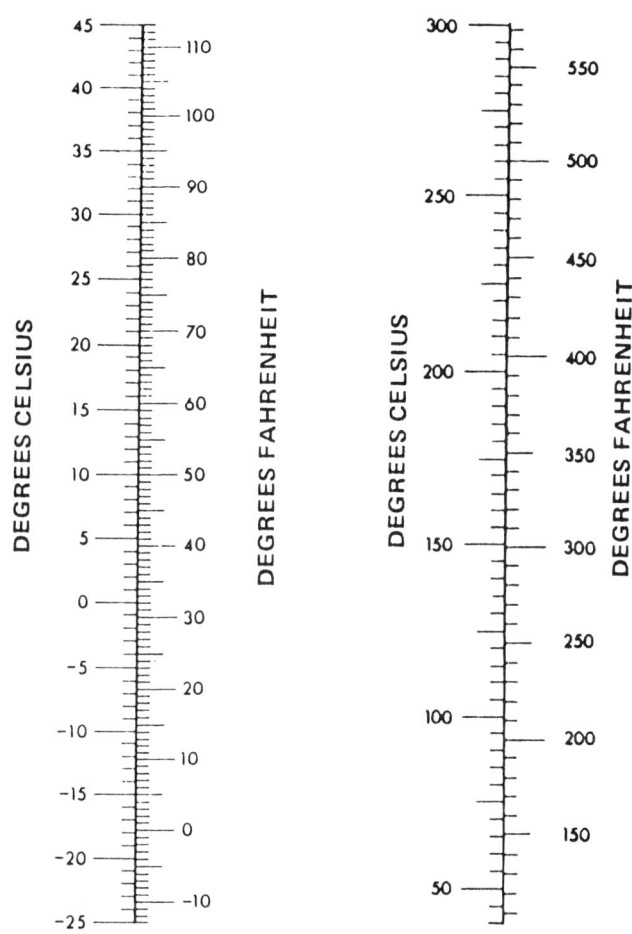

Source: John L. Frirer, *SI Metric Handbook*, New York: Charles Scribner's Sons, 1977, Figures 8-2, 8-3.

Appendix 20
The Use of Scientific Notation

Scientific notation is used to standardize and simplify writing numbers that have many significant digits. In Appendix 16, scientific notation is used in the multipliers that convert one unit of measurement to another. Scientific notation is indicated by the capital letter E, followed by a plus or negative sign, and a two digit integer.

The E signifies scientific notation and indicates that the multiplier that precedes the E will need to have the decimal place adjusted.

The *plus* or *negative sign* tells us the direction in which the decimal must be moved.

The *two digit number* following the plus or negative sign tells us how many places to move the decimal in the multiplier that precedes the E.

Example 1

The first example conversion taken from the AREA section of Appendix 22. To convert from ACRE to SQUARE METRE, Appendix 22 tells us to multiply by

$$4.046856\ E+03.$$

The $E+03$ tells us that we must move the decimal three places to the right. The multiplier becomes 4046.856. In effect, scientific notation has told us to multiply the 4.046856 by 10^3 (ten to the third power).

Ten acres is equivalent to 40,468.56 square metres (10 times $4046.856 = 40,468.56$). ∎

Example 2

The second example is also taken from the AREA section of Appendix 22. To convert from SQUARE METRE to SQUARE MILE, Appendix 22 tells us to multiply by

$$3.861022\ E\text{--}07.$$

The E–07 tells us that we must move the decimal seven places to the left. The multiplier becomes 0.0000003861022. In effect, scientific notation has told us to multiply the 3.861022 by 10^{-7} (ten to the minus seventh power).

Five thousand square metres is equivalent to 0.001930511 square miles (5000 times 0.0000003861022 = 0.001930511). ∎

Some of the conversions in Appendix 22, "Measurement Unit and Conversion Multipliers," include an *asterisk* (*) in the scientific notation between the E and the plus or negative sign, for example: 3.048000*E–01. In Appendix 22, the asterisk is used to inform us that this conversion is exact or sufficiently close to be considered exact. When using the multiplier to make the conversion, ignore the asterisk in your calculations. For example, 3.048000*E–01 is used as a multiplier the same as is 3.048000 E–01, i.e., by multiplying by 0.3048. Note that this use of an asterisk in scientific notation is not common.

Appendix 21
Prefixes for Metric System Multiples and Submultiples

Symbol	Name	Value
T	tera	10^{12}
G	giga	10^{9}
M	mega	10^{6}
my	myria	10^{4}
k	kilo	10^{3}
h	hecto	10^{2}
dk	deka	10
d	deci	10^{-1}
c	centi	10^{-2}
m	milli	10^{-3}
μ	micro	10^{-6}
n	nano	10^{-9}
p	pico	10^{-12}
f	femto	10^{-15}
a	atto	10^{-18}

Source: Random House, Inc., *The Random House Dictionary of the English Language*, Second Edition, Unabridged, 1987, p. 2225.

Appendix 22
Measurement Unit and Conversion Multipliers

ACCELERATION

Unit	Multiply by	to get	Multiply by	to get
foot per second per second (ft/s^2)	3.048 000*E-01	metre per second per second (m/s^2)	3.280 840 E+00	(ft/s^2)
inch per second per second (in/s^2)	2.540 000*E-02	metre per second per second (m/s^2)	3.937 008 E+01	(in/s^2)

ANGULAR MOMENTUM

Unit	Multiply by	to get	Multiply by	to get
pound foot squared per second ($lb \cdot ft^2/s$)	4.214 011 E-02	kilogram metre squared per second ($kg \cdot m^2/s$)	2.373 036 E+01	($lb \cdot ft^2/s$)
pound inch squared per second ($lb \cdot in^2/s$)	2.926 397 E-04	kilogram metre squared per second ($kg \cdot m^2/s$)	3.417 171 E+03	($lb \cdot in^2/s$)
ounce inch squared per second ($oz \cdot in^2/s$)	1.828 998 E-05	kilogram metre squared per second ($kg \cdot m^2/s$)	5.467 475 E+04	($oz \cdot in^2/s$)

AREA

Unit	Multiply by	to get	Multiply by	to get
acre	4.046 856 E+03	square metre (m^2)	2.471 054 E-04	acre
are	1.000 000*E+02	square metre (m^2)	1.000 000*E-02	(a)
square foot (ft^2)	9.290 304*E-02	square metre (m^2)	1.076 391 E+01	(ft^2)
square inch (in^2)	6.451 600*E-04	square metre (m^2)	1.550 003 E+03	(in^2)
square mile ($mile^2$)	2.589 988 E+06	square metre (m^2)	3.861 022 E-07	($mile^2$)
square rod (rod^2)	2.529 285 E+01	square metre (m^2)	3.953 686 E-02	(rod^2)
square yard (yd^2)	8.361 274 E-01	square metre (m^2)	1.195 990 E+00	(yd^2)

BENDING MOMENT AND TORQUE

Unit	Multiply by	to get	Multiply by	to get
dyne centimetre ($dyn \cdot cm$)	1.000 000*E-07	newton metre ($N \cdot m$)	1.000 000*E+07	($dyn \cdot cm$)
kilogram-force metre ($kgf \cdot m$)	9.806 650*E+00	newton metre ($N \cdot m$)	1.019 716 E-01	($kgf \cdot m$)
ounce-force inch ($ozf \cdot in$)	7.061 552 E-03	newton metre ($N \cdot m$)	1.416 119 E+02	($ozf \cdot in$)
pound-force inch ($lbf \cdot in$)	1.129 848 E-01	newton metre ($N \cdot m$)	8.850 748 E+00	($lbf \cdot in$)
pound-force foot ($lbf \cdot ft$)	1.355 818 E+00	newton metre ($N \cdot m$)	7.375 621 E-01	($lbf \cdot ft$)

Source: John L. Frirer, *SI Metric Handbook*, New York: Charles Scribner's Sons, 1977, Figure 12-9.

(BENDING MOMENT AND TORQUE) PER LENGTH

pound-force foot per inch (lbf·ft/in)	5.337 866 E+01	newton metre per metre (N·m/m)	1.873 408 E-02 (lbf·ft/in)
pound-force inch per inch (lbf·in/in)	4.448 222 E+00	newton metre per metre (N·m/m)	2.248 089 E-01 (lbf·in/in)

CAPACITY (See Volume)

DENSITY (See Mass per Volume)

ELECTRICITY AND MAGNETISM

ampere hour (A·h)	3.600 000*E+03	coulomb (C)	2.777 778 E-04 (A·h)
gauss (Gs)	1.000 000*E-04	tesla (T)	1.000 000*E+04 (Gs)
maxwell (Mx)	1.000 000*E-08	weber (Wb)	1.000 000*E+08 (Mx)
mho	1.000 000*E+00	siemens (S)	1.000 000*E+00 (mho)
oersted (Oe)	7.957 747 E+01	ampere per metre (A/m)	1.256 637 E-02 (Oe)

ENERGY, WORK
(Btu and calorie are International Table)

British thermal unit (Btu)	1.055 056 E+03	joule (J)	9.478 170 E-04 (Btu)
calorie (cal)	4.186 800*E+00	joule (J)	2.388 460 E-01 (cal)
electron volt (eV)	1.602 19 E-19	joule (J)	6.241 46 E+18 (eV)
erg (erg)	1.000 000*E-07	joule (J)	1.000 000*E+07 (erg)
foot pound-force (ft·lbf)	1.355 818 E+00	joule (J)	7.375 621 E-01 (ft·lbf)
foot poundal (ft·pdl)	4.214 011 E-02	joule (J)	2.373 036 E+01 (ft·pdl)
kilocalorie (kcal)	4.186 800*E+03	joule (J)	2.388 459 E-04 (kcal)
kilowatt hour (kW·h)	3.600 000*E+06	joule (J)	2.777 778 E-07 (kW·h)
watt hour (W·h)	3.600 000*E+03	joule (J)	2.777 778 E-04 (W·h)
watt second (W·s)	1.000 000*E+00	joule (J)	1.000 000*E+00 (W·s)

FLOW (See Mass per Time or Volume per Time)

FORCE

Unit	Multiply by	to get
dyne (dyn)	1.000 000*E-05	newton (N)
kilogram force (kgf)	9.806 650*E+00	newton (N)
kilopond (kp)	9.806 650*E+00	newton (N)
ounce force (ozf)	2.780 139 E-01	newton (N)
pound force (lbf)	4.448 222 E+00	newton (N)
poundal (pdl)	1.382 550 E-01	newton (N)

FORCE PER AREA (See Pressure)

FORCE PER LENGTH

Unit	Multiply by	to get
pound-force per inch (lbf/in)	1.751 268 E+02	newton per metre (N/m)
pound-force per foot (lbf/ft)	1.459 390 E+01	newton per metre (N/m)

HEAT
(Btu and calorie are International Table)

Unit	Multiply by	to get
Btu/(h·ft·°F), k, thermal conductivity	1.730 735 E+00	watt per metre kelvin W/(m·K)
Btu·in/(s·ft²·°F), k, thermal conductivity	5.192 204 E+02	watt per metre kelvin W/(m·K)
Btu·in/(h·ft²·°F), k, thermal conductivity	1.442 279 E-01	watt per metre kelvin W/(m·K)
Btu/ft²	1.135 653 E+04	joule per square metre (J/m²)
Btu/(h·ft²·°F), C, thermal conductance	5.678 263 E+00	watt per square metre kelvin W/(m²·K)
Btu/lb	2.326 000*E+03	joule per kilogram (J/kg)

Unit	Multiply by	to get
pound-force per inch (lbf/in)	5.710 148 E-03	(lbf/in)
pound-force per foot (lbf/ft)	6.852 178 E-02	(lbf/ft)
dyne (dyn)	1.000 000*E+05	(dyn)
kilogram force (kgf)	1.019 716 E-01	(kgf)
kilopond (kp)	1.019 716 E-01	(kp)
ounce force (ozf)	3.596 942 E+00	(ozf)
pound force (lbf)	2.248 089 E-01	(lbf)
poundal (pdl)	7.233 011 E+00	(pdl)
Btu/(h·ft·°F)	5.777 893 E-01	Btu/(h·ft·°F)
Btu·in/(s·ft²·°F)	1.925 964 E-03	Btu·in/(s·ft²·°F)
Btu·in/(h·ft²·°F)	6.933 471 E+00	Btu·in/(h·ft²·°F)
Btu/ft²	8.805 507 E-05	Btu/ft²
Btu/(h·ft²·°F)	1.761 102 E-01	Btu/(h·ft²·°F)
Btu/lb	4.299 226 E-04	Btu/lb

HEAT
(Btu and calorie are International Table)

Btu/(lb·°F)	4.186 800*E+03	joule per kilogram kelvin J/(kg·K)	2.388 459 E-04	Btu/(lb·°F)
c, heat capacity				
Btu/(s·ft²·°F)	2.044 175 E+04	watt per square metre kelvin W/(m²·K)	4.891 949 E-05	Btu/(s·ft²·°F)
C, thermal conductance				
cal/g	4.186 800*E+03	joule per kilogram (J/kg)	2.388 459 E-04	cal/g
cal/(g·°C)	4.186 800*E+03	joule per kilogram kelvin J/(kg·K)	2.388 459 E-04	cal/(g·°C)
c, heat capacity				
°F·h·ft²/Btu	1.761 102 E-01	kelvin square metre per watt K·m²/W	5.678 263 E+00	°F·h·ft²/Btu
R, thermal resistance				

LENGTH

foot (ft)	3.048 000*E-01	metre (m)	3.280 840 E+00	(ft)
inch (in)	2.540 000*E-02	metre (m)	3.937 008 E+01	(in)
microinch (µ in)	2.540 000*E-08	metre (m)	3.937 008 E+07	(µ in)
micron (µ)	1.000 000*E-06	metre (m)	1.000 000*E+06	(µ)
mil	2.540 000*E-05	metre (m)	3.937 008 E+04	mil
international nautical mile (n mile)	1.852 000*E+03	metre (m)	5.399 568 E-04	(n mile)
mile (mile)	1.609 344*E+03	metre (m)	6.213 712 E-04	(mile)
rod	5.029 200*E+00	metre (m)	1.988 388 E-01	rod
yard (yd)	9.144 000*E-01	metre (m)	1.093 613 E+00	(yd)

LIGHT

candle (cd) or candle power	1.000 000 E+00	candela (cd)	1.000 000 E+00	(cd)
foot-candle (fc)	1.076 391 E+01	lux (lx)	9.290 304 E-02	(fc)
foot-lambert (fL)	3.426 259 E+00	candela per square metre (cd/m²)	2.918 635 E-01	(fL)

Measurement Unit and Conversion Multipliers

MASS

Unit	Multiply by	to get
grain	6.479 891*E−05	kilogram (kg)
long hundredweight (cwt)	5.080 235 E+01	kilogram (kg)
short hundredweight (sh cwt)	4.535 924 E+01	kilogram (kg)
kilogram-force second squared per metre (kgf·s^2/m)	9.806 650*E+00	kilogram (kg)
ounce (oz)	2.834 952 E−02	kilogram (kg)
pennyweight (dwt)	1.555 174 E−03	kilogram (kg)
pound (lb)	4.535 924 E−01	kilogram (kg)
slug	1.459 390 E+01	kilogram (kg)
long ton (2240 lb)	1.016 047 E+03	kilogram (kg)
short ton (2000 lb)	9.071 847 E+02	kilogram (kg)
tonne (t)	1.000 000*E+03	kilogram (kg)

Unit	Multiply by	to get
kilogram (kg)	1.543 236 E+04	grain
kilogram (kg)	1.968 413 E−02	(cwt)
kilogram (kg)	2.204 622 E−02	(sh cwt)
kilogram (kg)	1.019 716 E−01	(kgf·s^2/m)
kilogram (kg)	3.527 397 E+01	(oz)
kilogram (kg)	6.430 149 E+02	(dwt)
kilogram (kg)	2.204 622 E+00	(lb)
kilogram (kg)	6.852 178 E−02	slug
kilogram (kg)	9.842 064 E−04	long ton (2240 lb)
kilogram (kg)	1.102 311 E−03	short ton (2000 lb)
kilogram (kg)	1.000 000*E−03	(t)

MASS PER AREA

Unit	Multiply by	to get
ounce per square yard (oz/yd^2)	3.390 575 E−02	kilogram per square metre (kg/m^2)
pound per square foot (lb/ft^2)	4.882 428 E+00	kilogram per square metre (kg/m^2)

Unit	Multiply by	to get
kilogram per square metre (kg/m^2)	2.949 352 E+01	(oz/yd^2)
kilogram per square metre (kg/m^2)	2.048 161 E−01	(lb/ft^2)

MASS PER CAPACITY (See Mass per Volume)

MASS PER TIME (Includes Flow)

Unit	Multiply by	to get
pound per second (lb/s)	4.535 924 E−01	kilogram per second (kg/s)
pound per minute (lb/min)	4.535 924 E−01	kilogram per minute (kg/min)
short ton per hour	9.071 847 E+02	kilogram per hour (kg/h)

Unit	Multiply by	to get
kilogram per second (kg/s)	2.204 622 E+00	(lb/s)
kilogram per minute (kg/min)	2.204 622 E+00	(lb/min)
kilogram per hour (kg/h)	1.102 311 E−03	short ton per hour

MASS PER VOLUME (Includes Density and Mass Capacity)

ounce per UK gallon (oz/UK gal)	6.236 021 E+00	kilogram per cubic metre (kg/m³)	1.603 586 E-01 (oz/UK gal)
ounce per UK gallon (oz/UK gal)	6.236 021 E-03	kilogram per litre (kg/l)	1.603 586 E+02 (oz/UK gal)
ounce per US gallon (oz/US gal)	7.489 152 E+00	kilogram per cubic metre (kg/m³)	1.335 265 E-01 (oz/US gal)
ounce per US gallon (oz/US gal)	7.489 152 E-03	kilogram per litre (kg/l)	1.335 265 E+02 (oz/US gal)
ounce per cubic inch (oz/in³)	1.729 994 E+03	kilogram per cubic metre (kg/m³)	5.780 367 E-04 (oz/in³)
pound per cubic foot (lb/ft³)	1.601 846 E+01	kilogram per cubic metre (kg/m³)	6.242 797 E-02 (lb/ft³)
pound per cubic inch (lb/in³)	2.767 990 E+04	kilogram per cubic metre (kg/m³)	3.612 730 E-05 (lb/in³)
pound per cubic yard (lb/yd³)	5.932 763 E-01	kilogram per cubic metre (kg/m³)	1.685 555 E+00 (lb/yd³)
pound per UK gallon (lb/UK gal)	9.977 633 E+01	kilogram per cubic metre (kg/m³)	1.002 242 E-02 (lb/UK gal)
pound per UK gallon (lb/UK gal)	9.977 633 E-02	kilogram per litre (kg/l)	1.002 242 E+01 (lb/UK gal)
pound per US gallon (lb/US gal)	1.198 264 E+02	kilogram per cubic metre (kg/m³)	8.345 406 E-03 (lb/US gal)
pound per US gallon (lb/US gal)	1.198 264 E-01	kilogram per litre (kg/l)	8.345 406 E+00 (lb/US gal)
slug per cubic foot (slug/ft³)	5.153 788 E+02	kilogram per cubic metre (kg/m³)	1.940 320 E-03 (slug/ft³)
long ton per cubic yard	1.328 939 E+03	kilogram per cubic metre (kg/m³)	7.524 800 E-04 long ton per cubic yard

Measurement Unit and Conversion Multipliers

MOMENT OF INERTIA

Units	Multiply by	to get
pound foot squared (lb·ft²)	4.214 011 E−02	kilogram metre squared (kg·m²)
pound inch squared (lb·in²)	2.926 397 E−04	kilogram metre squared (kg·m²)
ounce inch squared (oz·in²)	1.828 998 E−05	kilogram metre squared (kg·m²)

MOMENT OF MOMENTUM (See Angular Momentum)

MOMENTUM

Units	Multiply by	to get
pound foot per second (lb·ft/s)	1.382 550 E−01	kilogram metre per second (kg·m/s)
pound inch per second (lb·in/s)	1.152 125 E−02	kilogram metre per second (kg·m/s)
ounce inch per second (oz·in/s)	7.200 778 E−04	kilogram metre per second (kg·m/s)

POWER
(Btu is International Table)

Units	Multiply by	to get
Btu per minute (Btu/min)	1.758 427 E+01	watt (W)
Btu per hour (Btu/h)	2.930 711 E−01	watt (W)
erg per second (erg/s)	1.000 000*E−07	watt (W)
foot pound-force per hour (ft·lbf/h)	3.766 161 E−04	watt (W)
foot pound-force per minute (ft·lbf/min)	2.259 697 E−02	watt (W)
foot pound-force per second (ft·lbf/s)	1.355 818 E+00	watt (W)

MOMENT OF INERTIA

Units	Multiply by	to get
	2.373 036 E+01	(lb·ft²)
	3.417 171 E+03	(lb·in²)
	5.467 475 E+04	(oz·in²)

MOMENTUM

Units	Multiply by	to get
	7.233 011 E+00	(lb·ft/s)
	8.679 614 E+01	(lb·in/s)
	1.388 739 E+03	(oz·in/s)

POWER

Units	Multiply by	to get
	5.686 902 E−02	(Btu/min)
	3.412 141 E+00	(Btu/h)
	1.000 000*E+07	(erg/s)
	2.655 224 E+03	(ft·lbf/h)
	4.425 372 E+01	(ft·lbf/min)
	7.375 621 E−01	(ft·lbf/s)

POWER
(Btu is International Table)

horsepower, 550 foot pound-force per second (hp)	7.456 999 E+02	watt (W)	1.341 022 E-03	(hp)
electric horsepower	7.460 000*E+02	watt (W)	1.340 483 E-03	electric horsepower
metric horsepower in Germany: Pferdestärke (PS) in France: cheval vapeur (CV)	7.354 99 E+02	watt (W)	1.359 62 E-03	metric horsepower
kilogram-force metre per second (kgf·m/s)	9.806 65 E+00	watt (W)	1.019 72 E-01	(kgf·m/s)

PRESSURE OR STRESS (Force per Area)

normal atmosphere (atm)	1.013 25 E+05	pascal (Pa)	9.869 23 E-06	(atm)
bar	1.000 000*E+05	pascal (Pa)	1.000 000*E-05	bar
centimetre of mercury, 0°C (cm Hg)	1.333 22 E+03	pascal (Pa)	7.500 64 E-04	(cm Hg)
centimetre of water, 4°C (cm H_2O)	9.806 38 E+01	pascal (Pa)	1.019 74 E-02	(cm H_2O)
dyne per square centimetre (dyn/cm^2)	1.000 000*E-01	pascal (Pa)	1.000 000*E+01	(dyn/cm^2)
foot of water, 39.2°F (ft H_2O)	2.988 98 E+03	pascal (Pa)	3.345 62 E-04	(ft H_2O)
gram-force per square centimetre (gf/cm^2)	9.806 650*E+01	pascal (Pa)	1.019 716 E-02	(gf/cm^2)
inch of mercury, 60°F (in Hg)	3.376 85 E+03	pascal (Pa)	2.961 34 E-04	(in Hg)
inch of water, 60°F (in H_2O)	2.488 4 E+02	pascal (Pa)	4.018 6 E-03	(in H_2O)
kilogram-force per square centimetre (kgf/cm^2)	9.806 650*E+04	pascal (Pa)	1.019 716 E-05	(kgf/cm^2)
kilogram-force per square metre (kgf/m^2)	9.806 650*E+00	pascal (Pa)	1.019 716 E-01	(kgf/m^2)
kilogram-force per square millimetre (kgf/mm^2)	9.806 650*E+06	pascal (Pa)	1.019 716 E-07	(kgf/mm^2)
millimetre of mercury, 0°C (mm Hg)	1.333 224 E+02	pascal (Pa)	7.500 615 E-03	(mm Hg)

Measurement Unit and Conversion Multipliers

PRESSURE OR STRESS (Force per Area) (Continued)

Units	Multiply by	to get	Multiply by	to get
poundal per square foot (pdl/ft^2)	1.488 164 E+00	pascal (Pa)	6.719 689 E-01	(pdl/ft^2)
pound-force per square foot (lbf/ft^2)	4.788 026 E+01	pascal (Pa)	2.088 543 E-02	(lbf/ft^2)
pound-force per square inch, psi (lbf/in^2)	6.894 757 E+03	pascal (Pa)	1.450 377 E-04	(lbf/in^2)
torr	1.333 22 E+02	pascal (Pa)	7.500 62 E-03	torr

SPECIFIC VOLUME (See Volume per Mass)

SPEED (See Velocity)

STRESS (See Pressure)

SURFACE TENSION (See Force per Length)

TEMPERATURE

Units	Multiply by	to get	Multiply by	to get
degree Celsius (°C)	$t_K = t_C + 273.15$	kelvin (K)	$t_C = t_K - 273.15$	(°C)
degree Fahrenheit (°F)	$t_K = (t_F + 459.67)/1.8$	kelvin (K)	$t_F = 1.8 t_K - 459.67$	(°F)
degree Rankine (°R)	$t_K = t_R/1.8$	kelvin (K)	$t_R = 1.8 t_K$	(°R)
degree Fahrenheit (°F)	$t_C = (t_F - 32)/1.8$	degree Celsius (°C)	$t_F = 1.8 t_C + 32$	(°F)

TEMPERATURE INTERVAL

Units	Multiply by	to get	Multiply by	to get
degree Celsius (°C)	$t_K = t_C$	kelvin (K)	$t_C = t_K$	(°C)
degree Fahrenheit (°F)	$t_K = t_F/1.8$	kelvin (K)	$t_F = 1.8 t_K$	(°F)
degree Rankine (°R)	$t_K = t_R/1.8$	kelvin (K)	$t_R = 1.8 t_K$	(°R)
degree Fahrenheit (°F)	$t_C = t_F/1.8$	kelvin (K)	$t_F = 1.8 t_C$	(°F)

TIME

Units	Multiply by	to get	Multiply by	to get
day (d)	8.640 000 E+04	second (s)	1.157 407 E-05	(d)
hour (h)	3.600 000 E+03	second (s)	2.777 778 E-04	(h)
minute (min)	6.000 000 E+01	second (s)	1.666 667 E-02	(min)
month	2.628 000 E+06	second (s)	3.805 175 E-07	month
year (a)	3.153 600 E+07	second (s)	3.170 979 E-08	(a)

TORQUE (See Bending Moment)

VELOCITY (Includes Speed)

foot per hour (ft/h)	4.466 667 E−05	metre per second (m/s)	
foot per hour (ft/h)	3.048 000*E−01	metre per hour (m/h)	
foot per minute (ft/min)	5.080 000*E−03	metre per second (m/s)	
foot per minute (ft/min)	3.048 000*E−01	metre per minute (m/min)	
foot per second (ft/s)	3.048 000*E−01	metre per second (m/s)	
inch per second (in/s)	2.540 000*E−02	metre per second (m/s)	
kilometre per hour (km/h)	2.777 778 E−01	metre per second (m/s)	
international knot	1.852 000*E+00	kilometre per hour (km/h)	
mile per hour (mile/h)	4.470 400*E−01	metre per second (m/s)	
mile per hour (mile/h)	1.609 344*E+00	kilometre per hour (km/h)	
mile per minute (mile/min)	2.682 240*E+01	metre per second (m/s)	
mile per minute (mile/min)	1.609 344*E+03	metre per minute (m/min)	
mile per second (mile/s)	1.609 344*E+03	metre per second (m/s)	

	2.238 806 E−04		(ft/h)
	3.280 840 E+00		(ft/h)
	1.968 504 E+02		(ft/min)
	3.280 840 E+00		(ft/min)
	3.280 840 E+00		(ft/s)
	3.937 008 E+01		(in/s)
	3.600 000 E+00		(km/h)
	5.399 568 E−01		international knot
	2.236 936 E+00		(mile/h)
	6.213 712 E−01		(mile/h)
	3.728 227 E−02		(mile/min)
	6.213 712 E−04		(mile/min)
	6.213 712 E−04		(mile/s)

VISCOSITY

centipoise (cP)	1.000 000*E−03	pascal second (Pa·s)	
centistokes (cSt)	1.000 000*E−06	square metre per second (m²/s)	
poise (P)	1.000 000*E−01	pascal second (Pa·s)	
poundal second per square foot (pdl·s/ft²)	1.488 164 E+00	pascal second (Pa·s)	
pound per foot second (lb/ft·s)	1.488 164 E+00	pascal second (Pa·s)	
pound-force second per square foot (lbf·s/ft²)	4.788 026 E+01	pascal second (Pa·s)	
slug per foot second (slug/ft·s)	4.788 026 E+01	pascal second (Pa·s)	
square foot per second (ft²/s)	9.290 304*E−02	square metre per second (m²/s)	
stokes (St)	1.000 000*E−04	square metre per second (m²/s)	

	1.000 000*E+03		(cP)
	1.000 000*E+06		(cSt)
	1.000 000*E+01		(P)
	6.719 689 E−01		(pdl·s/ft²)
	6.719 689 E−01		(lb/ft·s)
	2.088 543 E−02		(lbf·s/ft²)
	2.088 543 E−02		(slug/ft·s)
	1.076 391 E+01		(ft²/s)
	1.000 000*E+04		(St)

Measurement Unit and Conversion Multipliers

VOLUME (Includes Capacity)

Unit	Multiply by	to get	Multiply by	to get
oil barrel (42 US gal)	1.589 873 E-01	cubic metre (m^3)	6.289 811 E+00	oil barrel
oil barrel (42 US gal)	1.589 873 E+02	litre (l)	6.289 811 E-03	oil barrel
US bushel	3.523 907 E-02	cubic metre (m^3)	2.837 759 E+01	US bushel
cubic foot (ft^3)	2.831 685 E-02	cubic metre (m^3)	3.531 466 E+01	(ft^3)
cubic inch (in^3)	1.638 706 E-05	cubic metre (m^3)	6.102 376 E+04	(in^3)
cubic inch (in^3)	1.638 706 E-02	litre (l)	6.102 376 E+01	(in^3)
cubic yard (yd^3)	7.645 549 E-01	cubic metre (m^3)	1.307 951 E+00	(yd^3)
cup	2.365 882 E-04	cubic metre (m^3)	4.226 753 E+03	cup
US fluid ounce (US fl oz)	2.957 353 E-05	cubic metre (m^3)	3.381 402 E+04	(US fl oz)
US fluid ounce (US fl oz)	2.957 353 E-02	litre (l)	3.381 402 E+01	(US fl oz)
Canadian liquid gallon	4.546 090 E-03	cubic metre (m^3)	2.199 692 E+02	Canadian liquid gallon
Canadian liquid gallon	4.546 090 E+00	litre (l)	2.199 692 E-01	Canadian liquid gallon
UK liquid gallon (UK gal)	4.546 092 E-03	cubic metre (m^3)	2.199 692 E+02	(UK gal)
UK liquid gallon (UK gal)	4.546 092 E+00	litre (l)	2.199 692 E-01	(UK gal)
US dry gallon (US dry gal)	4.404 884 E-03	cubic metre (m^3)	2.270 207 E+02	(US dry gal)
US liquid gallon (US gal)	3.785 412 E-03	cubic metre (m^3)	2.641 720 E+02	(US gal)
US liquid gallon (US gal)	3.785 412 E+00	litre (l)	2.641 720 E-01	(US gal)
UK fluid ounce (UK fl oz)	2.841 307 E-05	cubic metre (m^3)	3.519 507 E+04	(UK fl oz)
UK fluid ounce (UK fl oz)	2.841 307 E-02	litre (l)	3.519 507 E+01	(UK fl oz)
US peck	8.809 768 E-03	cubic metre (m^3)	1.135 104 E+02	US peck
US dry pint	5.506 105 E-04	cubic metre (m^3)	1.816 166 E+03	US dry pint
US liquid pint	4.731 765 E-04	cubic metre (m^3)	2.113 376 E+03	US liquid pint
US liquid pint	4.731 765 E-01	litre (l)	2.113 376 E+00	US liquid pint
US dry quart	1.101 221 E-03	cubic metre (m^3)	9.080 829 E+02	US dry quart
US liquid quart	9.463 529 E-04	cubic metre (m^3)	1.056 688 E+03	US liquid quart
US liquid quart	9.463 529 E-01	litre (l)	1.056 688 E+00	US liquid quart
tablespoon	1.478 676 E-05	cubic metre (m^3)	6.762 807 E+04	tablespoon
teaspoon	4.928 922 E-06	cubic metre (m^3)	2.028 841 E+05	teaspoon

VOLUME PER MASS

cubic foot per pound (ft^3/lb)	6.242 796 E-02	cubic metre per kilogram (m^3/kg)	1.601 846 E+01 (ft^3/lb)
cubic inch per pound (in^3/lb)	3.612 728 E-05	cubic metre per kilogram (m^3/kg)	2.767 991 E+04 (in^3/lb)

VOLUME PER TIME (Includes Flow)

cubic foot per minute (ft^3/min)	4.719 474 E-04	cubic metre per second (m^3/s)	2.118 880 E+03 (ft^3/min)
cubic foot per minute (ft^3/min)	2.831 685 E-02	cubic metre per minute (m^3/min)	3.531 466 E+01 (ft^3/min)
cubic foot per second (ft^3/s)	2.831 685 E-02	cubic metre per second (m^3/s)	3.531 466 E+01 (ft^3/s)
cubic inch per minute (in^3/min)	2.731 177 E-07	cubic metre per second (m^3/s)	3.661 425 E+06 (in^3/min)
cubic inch per minute (in^3/min)	1.638 706 E-05	cubic metre per minute (m^3/min)	6.102 375 E+04 (in^3/min)
cubic yard per minute (yd^3/min)	1.274 258 E-02	cubic metre per second (m^3/s)	7.847 704 E+01 (yd^3/min)
cubic yard per minute (yd^3/min)	7.645 549 E-01	cubic metre per minute (m^3/min)	1.307 950 E+00 (yd^3/min)
US liquid gallon per minute (US gal/min)	6.309 020 E-05	cubic metre per second (m^3/s)	1.585 032 E+04 (US gal/min)
US liquid gallon per minute (US gal/min)	6.309 020 E-02	litre per second (l/s)	1.585 032 E+01 (US gal/min)
US liquid gallon per minute (US gal/min)	3.785 412 E+00	litre per minute (l/min)	2.641 720 E-01 (US gal/min)
US liquid gallon per second (US gal/s)	3.785 412 E-03	cubic metre per second (m^3/s)	2.641 720 E+02 (US gal/s)
US liquid gallon per second (US gal/s)	3.785 412 E+00	litre per second (l/s)	2.641 720 E-01 (US gal/s)

Appendix 23
Consumer Price Index

Consumer Price Indexes for All Urban Consumers and for Urban Wage Earners and Clerical Workers: U.S. city average, by expenditure category and commodity or service group

(1982–84 = 100, unless otherwise indicated)

Series	Annual average 1988	Annual average 1989	1989 Aug.	1989 Sept.	1989 Oct.	1989 Nov.	1989 Dec.	1990 Jan.	1990 Feb.	1990 Mar.	1990 Apr.	1990 May	1990 June	1990 July	1990 Aug.
CONSUMER PRICE INDEX FOR ALL URBAN CONSUMERS:															
All items	118.3	124.0	124.6	125.0	125.6	125.9	126.1	127.4	128.0	128.7	128.9	129.2	129.9	130.4	131.6
All items (1967=100)	354.3	371.3	373.1	374.6	376.2	377.0	377.6	381.5	383.3	365.5	368.2	368.9	389.1	390.7	394.1
Food and beverages	118.2	124.9	125.6	125.9	126.3	126.7	127.2	130.0	130.9	131.2	131.0	131.1	131.7	132.4	132.7
Food	118.2	125.1	125.8	126.1	126.5	126.9	127.4	130.4	131.3	131.5	131.3	131.3	132.0	132.7	132.9
Food at home	116.6	124.2	124.9	125.0	125.4	125.8	126.5	131.0	132.1	131.9	131.1	130.9	131.7	132.5	132.7
Cereals and bakery products	122.1	132.4	134.1	134.6	135.0	135.3	136.1	136.9	137.4	137.6	138.9	139.3	140.1	140.5	141.4
Meats, poultry, fish, and eggs	114.3	121.3	122.3	122.9	122.4	122.8	123.8	126.2	126.7	127.9	128.2	127.8	129.9	130.4	131.1
Dairy products	108.4	115.6	114.5	116.1	118.2	120.2	122.9	125.8	126.9	126.8	125.2	124.7	124.9	125.7	127.3
Fruits and vegetables	126.1	138.0	138.8	136.6	137.1	137.8	136.7	153.7	157.9	153.9	149.0	147.4	147.1	149.4	146.1
Other foods at home	113.1	119.1	119.7	119.7	120.3	119.9	120.1	121.3	121.9	122.2	122.2	122.6	123.1	123.5	124.3
Sugar and sweets	114.0	119.4	120.6	120.8	121.3	120.7	121.1	122.5	122.9	123.0	123.F	124.4	124.5	124.9	125.6
Fats and oils	113.1	121.2	121.7	121.3	121.6	121.0	121.6	123.5	123.4	124.2	124.3	125.0	125.5	126.6	127.4
Nonalcoholic beverages	107.5	111.3	111.2	111.0	111.8	111.2	111.0	112.4	113.3	113.1	112.4	112.7	113.3	114.0	114.3
Other prepared foods	118.0	125.5	126.7	126.7	127.2	127.3	127.8	128.3	128.9	129.6	129.9	130.4	130.9	130.9	132.0
Food away from home	121.8	127.4	128.1	128.8	129.1	129.5	129.8	130.3	131.0	131.8	132.5	133.0	133.4	133.9	134.3
Alcoholic beverages	118.6	123.5	124.5	124.8	125.2	125.6	125.6	126.2	126.9	127.8	128.2	128.9	129.3	129.8	130.2
Housing	118.5	123.0	124.2	124.3	124.4	124.5	124.9	125.9	126.1	126.8	126.8	127.1	128.3	129.2	130.2
Shelter	127.1	132.8	134.1	134.1	134.8	135.2	135.6	136.3	136.6	137.8	138.0	138.3	139.5	141.1	142.4
Renters' costs (12/82=100)	133.6	138.9	141.5	139.4	140.0	140.1	140.1	142.0	143.5	144.8	144.7	144.3	145.3	148.7	150.7
Rent, residential	127.8	132.8	133.5	133.9	134.7	135.2	135.5	135.8	136.0	136.5	137.0	137.3	137.9	138.7	139.4
Other renters' costs	134.8	140.7	146.8	139.1	139.2	138.0	137.2	143.6	149.3	152.7	150.7	148.5	150.1	161.4	167.4
Homeowners' costs (12/82=100)	131.1	137.3	138.1	138.9	139.7	140.1	140.9	141.1	141.0	142.2	142.5	143.1	144.4	145.4	146.5
Owners' equivalent rent (12/82=100)	131.1	137.4	138.2	139.0	139.9	140.5	141.0	142.1	141.1	142.4	142.7	143.2	144.6	145.7	146.7
Household insurance (12/82=100)	129.0	132.6	133.3	133.6	133.7	133.8	134.0	134.1	134.5	134.8	134.4	134.9	135.2	135.3	135.6
Maintenance and repairs	114.7	118.0	118.5	118.6	118.6	119.3	119.5	120.4	120.8	121.2	121.2	122.2	121.8	122.1	121.2
Maintenance and repair services	117.9	120.6	121.3	120.9	121.0	121.7	122.2	123.7	124.8	124.8	125.6	126.2	125.4	125.6	124.1
Maintenance and repair commodities	110.4	114.6	114.8	115.6	115.5	116.2	115.8	116.0	115.9	116.4	115.4	116.7	117.0	117.4	117.5

Source: *Monthly Labor Review*, January 1991

(1982-84 = 100, unless otherwise indicated)

Series	Annual average 1988	Annual average 1989	1989 Aug.	1989 Sept.	1989 Oct.	1989 Nov.	1989 Dec.	1990 Jan.	1990 Feb.	1990 Mar.	1990 Apr.	1990 May	1990 June	1990 July	1990 Aug.
CONSUMER PRICE INDEX FOR ALL URBAN CONSUMERS:															
Fuel and other utilities	104.4	107.8	109.7	109.7	108.0	107.5	108.4	110.8	110.2	109.9	109.4	109.9	112.2	111.3	112.7
Fuels	98.0	100.9	103.7	103.5	101.0	99.9	101.2	104.5	103.1	102.3	101.2	101.9	105.4	104.5	105.6
Fuel oil, coal, and bottled gas	78.1	81.7	78.9	79.3	82.0	83.9	88.7	113.1	95.4	91.5	89.6	88.0	84.9	82.7	91.8
Gas (piped) and electricity	104.6	107.5	111.3	111.0	107.6	106.1	107.0	107.5	108.3	107.9	106.8	107.8	112.4	111.7	111.8
Other utilities and public services	122.9	127.1	127.8	128.1	127.6	127.9	128.2	129.3	130.0	130.7	130.9	131.2	131.8	130.8	132.8
Household furnishings and operations	109.4	111.2	111.4	111.7	111.9	111.9	111.7	112.1	112.8	112.8	112.8	113.2	113.1	113.6	113.3
Housefurnishings	105.1	105.5	105.2	105.7	106.1	106.0	105.5	106.1	106.9	106.9	106.6	106.7	106.3	106.8	106.5
Housekeeping supplies	114.7	120.9	122.3	122.3	122.5	122.5	123.6	123.2	123.5	123.4	123.9	125.0	125.8	125.9	125.6
Housekeeping services	114.3	117.3	117.5	117.5	117.4	117.6	117.6	117.9	118.4	118.7	119.1	119.5	119.8	120.5	120.4
Apparel and upkeep	115.4	118.6	115.0	120.0	122.7	122.1	119.2	116.7	120.4	125.4	126.7	125.5	123.3	120.8	122.2
Apparel commodities	113.7	116.7	112.8	118.2	121.1	120.4	117.1	114.3	118.3	123.7	125.0	123.6	121.1	118.4	119.0
Men's and boys' apparel	113.4	117.0	114.7	117.7	120.3	121.1	118.8	116.3	117.0	119.3	121.0	121.9	119.9	118.6	119.3
Women's and girls' apparel	114.9	116.4	109.5	119.0	123.1	121.3	116.4	112.0	117.7	126.8	127.9	124.7	120.9	116.1	118.9
Infants' and toddlers' apparel	116.4	119.1	116.7	118.0	118.3	117.2	115.3	112.7	124.3	130.0	130.0	127.2	127.8	127.7	126.5
Footwear	109.9	114.4	112.6	114.1	117.6	116.6	114.7	113.1	114.5	116.9	118.6	118.5	117.3	116.1	116.3
Other apparel commodities	116.0	122.1	124.1	124.5	123.0	123.5	122.8	125.1	130.6	132.7	132.8	132.1	131.4	131.1	131.3
Apparel services	123.7	129.4	129.5	129.7	129.8	130.8	131.3	132.4	132.9	133.6	134.8	136.2	136.4	136.8	138.2
Transportation	108.7	114.1	114.3	113.7	114.5	115.0	115.2	117.2	117.1	116.8	117.3	117.7	118.2	118.4	120.6
Private transportation	107.6	112.9	113.1	112.4	113.3	113.7	113.9	115.9	115.6	115.1	115.5	115.9	116.4	116.6	119.0
New vehicles	116.5	119.2	117.7	117.1	118.5	120.1	118.8	122.4	122.2	121.6	121.6	121.0	120.6	120.2	119.9
New cars	116.9	119.2	117.7	117.0	118.6	120.5	121.8	122.3	121.9	121.3	120.7	120.7	120.3	119.8	119.5
Used cars	118.0	120.4	120.3	119.8	119.7	120.1	119.7	118.9	117.4	116.6	116.2	116.2	117.6	118.2	118.3
Motor fuel	80.9	88.5	91.0	88.8	88.9	87.2	85.8	91.4	90.6	89.3	91.2	92.5	94.6	94.3	103.2
Gasoline	80.8	88.5	91.1	88.8	88.8	87.0	85.5	90.6	90.2	89.1	91.0	92.4	94.6	94.4	103.1
Maintenance and repair	119.7	124.9	125.4	126.2	126.7	126.7	126.9	127.3	127.6	128.8	129.4	129.4	129.6	130.2	130.4
Other private transportation	127.9	135.8	135.7	135.7	137.1	138.2	139.0	140.3	140.8	140.7	140.8	140.8	141.0	142.1	142.4
Other private transportation commodities	98.9	101.5	102.0	102.0	101.9	102.1	102.3	101.9	102.1	102.0	101.9	101.8	101.8	101.7	102.2
Other private transportation services	133.9	143.2	142.9	142.9	144.8	146.0	146.9	148.7	149.3	149.2	149.4	149.3	149.7	151.0	151.3
Public transportation	123.3	129.5	130.1	130.1	130.6	131.3	131.7	134.2	136.7	139.1	140.3	140.9	141.5	141.6	141.9

Appendix 23

Consumer Price Index

Medical care	138.6	149.3	150.7	151.7	152.7	153.9	154.4	155.9	157.5	158.7	159.8	160.8	161.9	163.5	165.0
Medical care commodities	139.9	150.8	152.1	153.3	154.1	155.3	156.0	156.9	158.6	159.9	161.3	162.2	163.3	164.1	164.8
Medical care services	138.3	148.9	150.4	151.3	152.3	153.6	154.1	155.7	157.2	158.5	159.4	160.5	161.5	163.4	165.0
Professional services	137.5	146.4	147.5	148.0	148.6	149.3	149.9	151.1	152.3	153.2	154.1	155.1	155.8	157.0	157.8
Hospital and related services	143.9	160.5	162.7	164.3	166.0	167.0	167.9	169.9	171.6	173.0	173.7	174.3	175.4	178.1	180.9
Entertainment	120.3	126.5	127.3	127.8	128.4	128.6	129.1	129.9	130.4	130.9	131.4	131.7	131.9	132.7	133.0
Entertainment commodities	115.0	119.8	120.0	120.5	121.2	121.3	121.6	122.3	122.5	123.1	123.5	123.7	123.5	124.4	124.8
Entertainment services	127.7	135.4	136.7	137.2	137.8	138.2	138.8	139.8	140.5	141.0	141.6	142.0	142.6	143.3	143.6
Other goods and services	137.0	147.7	148.7	151.2	151.8	151.9	152.9	154.0	154.7	155.2	155.8	156.6	157.8	159.2	160.4
Tobacco products	145.8	164.4	168.8	168.2	168.8	168.6	171.9	174.1	175.0	175.1	175.6	176.7	180.9	185.7	185.8
Personal care	119.4	125.0	125.6	125.9	126.4	127.0	127.1	127.6	128.4	129.0	130.3	130.2	131.0	130.8	130.6
Toilet goods and personal care appliances	118.1	123.2	123.8	124.0	124.4	125.1	124.7	125.1	126.0	126.9	128.3	128.3	129.2	128.4	128.1
Personal care services	120.7	126.8	127.3	127.7	128.5	129.0	129.7	130.3	130.9	131.2	132.3	132.1	132.8	132.9	133.3
Personal and educational expenses	147.9	158.1	158.1	162.9	163.5	163.5	164.0	165.1	165.8	166.3	166.6	167.7	168.0	168.9	171.2
School books and supplies	148.1	158.0	156.6	163.0	163.6	163.9	164.0	167.9	169.7	169.9	169.9	169.9	169.8	170.3	170.5
Personal and educational services	148.0	158.3	158.4	163.1	163.7	163.7	164.2	165.1	165.6	166.3	166.6	167.7	168.1	169.0	171.5
All items	118.3	124.0	124.6	125.0	125.6	125.9	126.1	127.4	128.0	128.7	128.9	129.2	129.9	130.4	131.6
Commodities	111.5	116.7	116.7	117.3	118.1	118.3	118.2	119.9	120.6	121.1	121.4	121.4	121.6	121.6	122.8
Food and beverages	118.2	124.9	125.6	125.9	126.3	126.7	127.2	130.0	130.9	130.9	131.0	131.1	131.7	132.4	132.7
Commodities less food and beverages	107.3	111.6	111.1	111.9	113.0	113.0	112.6	113.7	114.2	114.9	115.4	115.5	115.4	115.0	116.8
Nondurables less food and beverages	105.2	111.2	110.9	112.4	113.6	113.1	112.0	113.7	114.5	116.1	117.1	117.1	117.1	116.4	119.5
Apparel commodities	113.7	116.7	112.8	118.2	121.1	120.4	117.1	114.3	118.3	123.7	125.0	123.6	121.1	118.4	119.9
Nondurables less food, beverages, and apparel	103.2	111.0	112.5	112.0	112.4	111.9	112.0	116.0	115.3	114.8	115.7	116.5	117.7	118.1	122.1
Durables	110.4	112.2	111.4	111.3	112.1	113.0	113.5	113.8	113.7	113.4	113.1	113.2	112.9	113.0	112.9
Services	125.7	131.9	133.1	133.4	133.7	134.1	134.6	135.4	136.0	136.9	137.1	137.6	138.8	139.9	140.9
Rent of shelter (12/82=100)	132.0	138.0	139.3	139.3	140.1	140.5	141.6	142.0	142.0	143.3	143.5	143.7	145.0	146.7	149.1
Household services less rent of shelter (12/82=100)	115.3	118.7	120.7	120.7	119.0	118.5	119.0	119.6	120.3	120.5	120.1	120.8	123.1	122.6	123.2
Transportation services	128.0	135.6	135.7	135.9	137.1	138.0	138.6	140.2	141.1	141.9	142.4	142.5	142.9	143.8	144.0
Medical care services	138.3	148.9	150.4	151.3	152.3	153.6	154.1	155.7	157.2	158.5	159.4	160.5	161.5	163.4	165.0
Other services	132.6	140.9	141.5	143.8	144.3	144.6	145.1	146.1	146.5	147.2	147.8	148.5	148.9	149.6	151.0

1982-84=100, unless otherwise indicated)

Series	Annual average 1988	Annual average 1989	1989 Aug.	1989 Sept.	1989 Oct.	1989 Nov.	1989 Dec.	1990 Jan.	1990 Feb.	1990 Mar.	1990 Apr.	1990 May	1990 June	1990 July	1990 Aug.
Special indexes:															
All items less food	118.3	123.7	124.3	124.8	125.4	125.6	125.8	126.7	127.3	128.1	128.4	128.7	129.4	130.0	131.3
All items less shelter	115.9	121.6	122.0	122.6	123.1	123.3	123.5	125.0	125.7	126.2	126.5	126.7	127.3	127.5	128.6
All items less homeowners costs (12/82=100)	119.5	125.3	125.9	126.3	126.8	127.0	127.1	128.7	129.5	130.1	130.4	130.6	131.2	131.6	132.8
All items less medical care	117.0	122.4	123.0	123.4	124.0	124.2	124.4	125.7	126.2	126.9	127.1	127.3	128.0	128.5	129.6
Commodities less food	107.7	112.0	111.6	112.4	113.4	113.4	113.0	114.1	114.6	115.4	115.9	115.9	115.8	115.5	117.2
Nondurables less food	105.8	111.7	111.5	112.9	114.1	113.6	112.6	114.2	115.0	116.5	117.4	117.5	117.6	117.0	119.9
Nondurables less food and apparel	104.0	111.3	112.8	112.4	112.8	112.4	112.5	116.1	115.5	115.2	116.0	116.8	118.0	118.3	121.9
Nondurables	111.8	118.2	118.4	119.3	120.1	120.0	119.8	122.0	122.9	123.8	124.2	124.2	124.6	124.6	126.3
Services less rent of shelter (12/82=100)	128.3	135.1	136.3	137.0	137.0	137.2	137.8	138.9	139.8	140.3	140.6	141.2	142.5	143.0	143.8
Services less medical care	124.3	130.1	131.3	131.6	131.8	132.1	132.6	133.4	133.9	134.7	134.9	135.3	136.5	137.5	138.5
Energy	89.3	94.3	97.0	95.9	94.6	93.2	93.2	97.6	96.4	95.5	95.7	96.7	99.5	98.9	103.6
All items less energy	122.3	128.1	128.5	129.1	129.9	130.4	130.6	131.5	132.3	133.3	133.5	133.7	134.2	134.8	135.6
All items less food and energy	123.4	129.0	129.3	130.0	130.9	131.3	131.5	132.0	132.8	133.9	134.2	134.4	134.8	135.5	130.4
Commodities less food and energy	115.8	119.6	118.8	120.1	121.2	121.6	121.2	121.0	122.2	123.4	123.7	123.6	123.2	122.9	123.2
Energy commodities	80.8	87.9	89.8	88.0	88.3	87.0	86.4	94.2	91.3	89.8	91.2	92.2	93.7	93.2	102.1
Services less energy	127.9	134.4	135.4	135.8	136.5	137.0	137.5	138.4	138.9	140.0	140.7	140.7	141.6	142.8	144.0
Purchasing power of the consumer dollar:															
1982-84=$1.00	84.6	80.7	80.3	80.0	79.6	79.5	79.3	78.5	78.2	77.7	77.6	77.4	77.0	76.7	76.0
1967=$1.00	28.2	26.9	26.8	26.7	26.6	26.5	26.5	26.2	26.1	25.9	25.9	25.8	25.7	25.6	25.4
CONSUMER PRICE INDEX FOR URBAN WAGE EARNERS AND CLERICAL WORKERS:															
All items	117.0	122.6	123.2	123.6	124.2	124.4	124.6	125.9	126.4	127.1	127.3	127.5	128.3	128.7	129.9
All items (1967=100)	348.4	365.2	367.0	368.3	369.8	370.6	371.1	375.0	376.6	378.5	379.2	379.9	382.1	383.4	386.9

Consumer Price Index

Food and beverages	117.9	124.6	125.3	125.6	126.0	126.4	126.9	129.7	130.6	130.9	130.7	130.7	131.5	132.1	132.4
Food	117.9	124.8	125.5	125.8	126.0	126.6	127.1	130.1	131.1	131.2	130.9	131.0	131.8	132.4	132.7
Food at home	116.2	123.9	124.6	124.6	125.0	125.5	126.2	130.5	131.6	131.5	130.6	130.4	131.4	132.2	132.4
Cereals and bakery products	122.2	132.4	134.1	134.6	135.1	135.3	136.0	136.8	137.4	137.6	138.8	139.2	140.0	140.4	141.3
Meats, poultry, fish, and eggs	114.1	121.2	122.1	122.7	122.2	122.9	123.8	126.7	126.6	127.6	128.1	127.8	130.0	130.5	131.2
Dairy products	108.1	115.4	114.2	115.9	118.0	118.0	122.8	125.1	126.9	126.8	126.6	124.6	124.8	125.5	127.3
Fruits and vegetables	127.6	137.6	138.6	136.1	136.5	137.0	135.8	152.9	157.7	153.3	147.9	146.4	146.6	148.9	145.6
Other foods at home	113.0	119.0	119.6	119.6	120.2	119.8	120.1	121.3	121.8	122.2	122.1	122.6	123.1	123.5	124.2
Sugar and sweets	113.9	119.5	120.6	120.9	121.4	120.7	121.5	122.5	123.0	123.1	123.7	124.4	124.6	124.9	125.7
Fats and oils	113.0	121.1	121.6	121.2	121.5	120.9	121.5	123.4	123.2	124.0	124.1	124.9	125.4	126.4	127.3
Nonalcoholic beverages	107.7	111.4	111.1	111.0	112.0	111.3	111.2	112.7	113.6	113.4	112.7	112.9	113.6	114.2	114.6
Other prepared foods	117.8	125.3	126.5	126.6	127.0	127.1	127.4	128.2	128.7	129.5	129.7	130.2	130.8	130.7	131.8
Food away from home	121.6	127.3	128.0	128.6	129.0	129.4	129.7	130.2	130.9	131.7	132.3	132.8	133.2	133.7	134.1
Alcoholic beverages	118.3	123.1	124.0	124.4	124.7	125.1	125.2	125.9	126.7	127.4	128.0	128.7	129.1	129.5	129.8
Housing	116.8	121.2	122.4	122.5	122.5	122.7	123.1	123.9	124.1	124.7	124.7	125.1	126.2	127.0	127.9
Shelter	124.3	129.8	131.0	131.1	131.8	132.3	132.6	133.2	133.4	134.5	134.7	135.0	136.1	137.5	138.7
Renters' costs (12/84=100)	119.2	123.9	125.9	124.6	125.1	125.3	126.6	130.5	127.5	128.4	128.4	128.4	129.2	131.4	132.7
Rent, residential	127.5	132.3	133.0	133.4	134.2	134.6	135.0	135.3	135.4	136.0	136.4	136.8	137.4	138.2	138.8
Other renters' costs	135.2	141.5	152.0	140.9	140.4	139.1	144.1	144.1	149.8	153.2	150.9	148.8	150.7	161.9	167.9
Homeowners' costs (12/84=100)	119.5	125.1	125.8	126.6	127.3	127.6	128.3	128.5	128.5	129.6	129.9	130.3	131.5	132.4	133.5
Household insurance (12/84=100)	119.5	125.2	125.9	126.7	127.4	128.0	128.5	128.6	128.6	129.7	130.0	130.4	131.6	132.6	133.7
Maintenance and repairs	118.2	121.4	122.0	122.4	122.5	122.5	122.7	122.8	123.1	123.3	123.0	123.6	123.8	123.9	124.1
Maintenance and repair services	117.7	120.4	121.3	120.7	120.9	121.7	122.1	122.4	122.0	123.1	123.0	121.7	121.8	122.1	121.3
Maintenance and repair commodities	108.3	112.6	112.5	113.3	113.4	114.0	113.6	113.8	114.3	114.3	114.3	114.3	114.9	126.6	125.2
Fuel and other utilities	104.1	107.5	109.5	120.7	107.6	107.2	108.0	110.2	108.8	109.6	109.8	109.5	112.0	111.1	115.3
Fuels	97.7	100.6	103.5	103.3	99.5	107.2	100.7	103.8	102.5	101.8	100.6	109.0	105.0	104.2	112.4
Fuel oil, coal, and bottled gas	77.9	81.4	78.8	79.2	83.6	88.1	106.7	112.7	95.2	91.8	89.4	91.3	87.7	84.9	105.1
Gas (piped) and electricity	104.4	107.3	111.0	110.7	105.8	106.7	106.7	107.2	107.9	107.5	106.4	107.2	112.1	111.4	91.6
Other utilities and public services	112.9	127.4	128.0	128.3	127.8	128.4	128.4	129.6	130.4	131.0	131.4	131.7	132.3	131.2	113.3
Household furnishings and operations	108.9	110.6	110.8	111.0	111.2	111.1	111.5	112.1	112.1	112.1	112.2	112.4	112.3	112.7	133.3
Housefurnishings	104.5	104.8	104.6	105.0	105.3	105.2	104.7	105.3	106.1	105.9	105.8	105.8	105.3	105.8	105.6
Housekeeping supplies	115.1	121.2	122.6	122.6	122.7	122.7	123.8	123.5	123.8	123.9	124.4	125.3	126.1	126.2	125.8
Housekeeping services	115.0	117.4	117.6	117.6	117.5	117.7	117.8	118.1	118.7	119.0	119.3	119.7	119.9	120.4	120.4

(1982-84=100, unless otherwise indicated)

Series	Annual average 1988	Annual average 1989	1989 Aug.	1989 Sept.	1989 Oct.	1989 Nov.	1989 Dec.	1990 Jan.	1990 Feb.	1990 Mar.	1990 Apr.	1990 May	1990 June	1990 July	1990 Aug.
Apparel and upkeep	114.9	117.9	114.5	119.3	122.0	121.4	118.5	116.1	119.3	124.4	125.8	124.7	122.4	119.8	121.3
Apparel commodities	113.4	116.1	112.4	117.6	120.5	119.8	116.6	114.0	117.3	122.8	124.2	122.9	120.4	117.6	119.0
Men's and boys' apparel	112.8	116.1	113.9	116.9	119.6	120.2	118.0	115.8	116.2	118.3	120.0	120.7	118.9	117.4	118.0
Women's and girls' apparel	114.5	115.5	108.9	118.1	122.0	120.5	115.5	111.3	116.4	125.7	126.9	123.8	119.8	115.0	118.1
Infants' and toddlers' apparel	118.6	122.5	120.4	122.0	122.2	121.0	119.3	116.6	127.1	129.9	132.2	130.2	130.2	129.8	129.2
Footwear	110.4	114.7	113.1	114.5	118.0	117.0	115.4	113.8	115.0	117.4	119.2	119.3	118.3	116.9	116.8
Other apparel commodities	114.9	120.5	122.4	122.5	121.9	122.4	121.5	123.2	127.0	130.5	130.7	130.3	128.8	128.2	128.1
Apparel services	123.0	128.6	128.7	128.8	129.0	130.0	130.6	131.7	132.2	133.2	134.2	135.5	135.6	135.9	137.6
Transportation	108.3	113.9	114.2	113.5	114.3	114.6	114.8	116.8	116.6	116.2	116.6	117.1	117.7	117.8	120.3
Private transportation	107.5	113.0	113.3	112.6	113.3	113.7	113.8	115.8	115.5	114.9	115.4	115.8	116.4	116.5	119.1
New vehicles	116.2	119.0	117.6	117.1	118.4	120.5	122.0	122.4	122.3	121.7	121.2	121.1	120.7	120.3	120.0
New cars	116.6	119.1	117.6	116.9	118.4	120.2	121.7	122.2	121.8	121.2	120.6	120.5	120.2	119.7	119.3
Used cars	117.9	120.3	120.1	119.6	119.5	119.9	119.5	118.7	117.2	116.4	116.0	116.6	117.3	118.0	118.0
Motor fuel	80.9	88.6	91.0	89.0	89.1	87.3	85.9	91.7	90.7	89.4	91.3	92.6	94.7	94.4	103.4
Gasoline	80.8	88.6	91.2	89.0	89.0	87.2	85.6	91.0	90.4	89.2	91.2	92.5	94.8	94.5	103.3
Maintenance and repair	119.8	124.9	125.4	126.2	126.7	126.8	126.9	127.3	127.9	129.0	129.6	129.7	129.9	130.3	130.7
Other private transportation	125.8	133.7	133.7	133.6	134.9	136.0	136.8	138.1	138.5	138.3	138.4	138.6	138.6	139.5	139.7
Other private transportation commodities	98.6	101.1	101.8	101.6	101.5	101.7	101.9	101.4	101.7	101.5	101.4	101.3	101.3	101.3	101.7
Other private transportation services	131.7	141.0	140.8	140.6	142.5	143.8	144.7	146.5	146.9	146.8	146.8	146.8	147.2	148.5	148.5
Public transportation	122.5	128.2	129.1	129.1	129.4	129.7	130.1	132.9	135.4	137.4	138.4	138.9	139.6	139.7	140.0
Medical care	139.0	149.6	151.1	152.1	153.0	154.2	154.7	156.1	157.6	158.8	159.8	160.8	161.8	163.3	164.7
Medical care commodities	139.0	149.7	150.9	152.2	153.1	154.2	154.8	155.7	157.4	158.6	160.0	161.0	162.1	162.9	163.7
Medical care services	139.0	149.6	151.1	152.1	153.0	154.2	154.7	156.2	157.7	158.8	159.7	160.7	161.1	163.4	165.0
Professional services	137.7	146.7	147.8	148.4	149.0	149.6	150.2	151.5	152.6	153.5	154.3	155.3	156.1	157.2	158.1
Hospital and related services	143.3	159.4	161.6	163.3	164.7	166.5	166.8	168.4	170.1	171.3	172.1	172.7	173.8	176.3	178.8
Entertainment	119.7	125.8	126.5	127.0	127.7	127.9	128.4	129.1	129.5	130.0	130.6	130.8	131.0	131.7	132.1
Entertainment commodities	115.1	119.9	120.1	120.6	121.3	121.4	121.7	122.3	122.4	123.0	123.4	123.6	123.4	124.2	124.7
Entertainment services	127.2	135.1	136.4	137.1	137.6	138.0	138.7	139.6	140.4	140.9	141.6	141.9	142.5	143.1	143.4
Other goods and services	136.5	147.4	148.8	150.8	151.4	151.5	152.7	153.9	154.6	155.1	155.7	156.3	157.8	159.4	160.5
Tobacco products	146.0	164.2	168.5	168.0	168.6	168.5	171.8	173.8	174.8	174.8	175.3	176.4	180.6	185.4	185.5
Personal care	119.3	124.8	125.4	125.7	126.3	126.8	126.9	127.3	128.1	128.7	130.0	129.9	130.7	130.3	130.5
Toilet goods and personal care appliances	118.0	123.3	123.8	124.1	124.6	125.1	124.7	124.9	126.0	126.8	128.2	128.1	129.1	128.2	128.2

Consumer Price Index

Personal care services	120.5	126.6	127.1	127.5	128.2	128.7	129.4	130.1	130.5	130.8	132.1	131.9	132.6	132.8	133.2
Personal and educational expenses	147.4	157.3	157.3	161.8	162.5	162.5	163.1	164.2	164.8	165.6	166.0	166.5	166.9	167.7	169.9
School books and supplies	147.1	156.9	155.6	161.7	162.8	162.8	162.9	166.9	168.5	168.6	168.6	168.6	168.6	169.2	169.6
Personal and educational services	147.7	157.7	157.8	162.1	162.7	162.8	163.4	164.3	164.8	165.7	166.1	166.7	167.1	167.9	170.3
All items	117.0	122.6	123.2	123.6	124.2	124.4	124.6	125.9	126.4	127.1	127.3	127.5	128.3	128.7	129.9
Commodities	111.0	116.3	116.4	116.9	117.7	117.8	117.8	119.5	120.1	120.5	120.8	120.9	121.2	121.3	122.6
Food and beverages	117.9	124.6	125.3	125.6	126.0	126.4	126.9	129.7	130.6	130.9	130.7	130.7	131.5	132.1	132.4
Commodities less food and beverages	106.8	111.2	110.9	111.6	112.5	112.5	112.1	113.3	113.6	114.2	114.8	114.9	114.9	114.6	116.5
Nondurables less food and beverages	104.6	110.9	110.8	112.0	113.2	112.6	111.6	113.4	114.0	115.4	116.5	116.6	116.8	116.2	119.6
Apparel commodities	113.4	116.1	112.4	117.6	120.5	119.8	116.6	114.0	117.3	122.8	124.2	122.9	120.4	117.6	119.0
Nondurables less food, beverages, and apparel	102.9	110.9	112.6	112.0	112.3	111.7	111.7	115.7	115.0	114.5	115.5	116.3	117.8	118.2	122.6
Durables	108.9	110.8	110.1	110.0	110.6	111.6	112.0	112.2	112.0	111.6	111.4	111.1	111.2	111.4	111.3
Services	124.7	130.8	132.0	132.6	132.6	132.9	133.4	134.2	134.8	135.6	135.8	136.2	137.4	138.3	139.3
Rent of shelter (12/84=100)	119.4	125.9	125.9	126.0	126.7	127.1	127.5	128.0	129.3	129.3	129.5	130.8	132.2	133.4
Household services less rent of shelter (12/84=100)	105.9	109.1	111.0	111.0	109.3	108.8	109.3	110.0	110.6	110.7	110.3	110.9	113.3	112.7	113.3
Transportation services	127.1	134.8	134.9	135.0	136.3	137.1	137.8	139.4	140.2	140.7	141.1	141.2	141.5	142.4	142.5
Medical care services	139.0	149.6	151.1	152.1	153.0	154.2	154.7	156.2	157.7	158.8	159.7	160.7	161.7	163.4	165.0
Other services	131.4	139.6	140.1	142.3	142.9	143.2	143.8	144.7	145.3	145.9	146.6	147.1	147.5	148.1	149.4
Special indexes:																
All items less food	116.7	122.0	122.6	123.1	123.6	123.8	124.0	124.9	125.3	126.1	126.4	126.7	127.4	127.8	129.2
All items less shelter	115.2	120.9	121.3	121.8	122.3	122.5	122.6	124.2	124.8	125.3	125.5	125.8	126.4	126.5	127.7
All items less homeowners' costs (12/84=100)	110.4	115.7	116.3	116.6	117.1	117.3	117.4	118.8	119.4	119.9	119.9	120.3	121.0	121.3	122.4
All items less medical care	115.8	121.2	121.8	122.2	122.7	122.9	123.1	124.4	124.9	125.5	125.7	125.9	126.6	127.0	128.2
Commodities less food	107.2	111.6	111.4	112.1	112.9	112.9	112.6	113.7	114.0	114.6	115.2	115.3	115.4	115.1	117.0
Nondurables less food	110.4	111.3	111.4	111.6	112.5	113.1	113.7	113.9	114.5	115.8	116.9	117.1	117.3	116.8	119.9
Nondurables less food and apparel	103.7	111.2	112.8	112.3	112.7	112.1	112.2	113.9	115.3	114.9	115.8	116.7	118.0	118.3	122.3
Nondurables	111.5	118.0	118.3	119.1	119.8	119.7	119.5	115.8	121.8	122.6	123.4	123.8	124.4	124.4	126.3
Services less rent of shelter (12/84=100)	115.6	121.7	122.7	123.3	123.2	123.4	123.9	124.9	125.7	126.1	126.3	126.8	128.0	128.4	129.1
Services less medical care	123.3	129.0	130.1	130.4	130.6	130.9	131.4	132.2	132.7	133.4	133.6	133.9	135.1	136.0	136.9
Energy	88.6	93.9	96.6	95.5	94.2	92.8	92.7	97.1	96.0	94.9	95.4	96.3	99.2	98.7	103.7
All items less energy	121.0	126.7	127.1	127.7	128.5	128.9	129.1	130.1	130.7	131.6	131.9	132.0	132.5	133.1	133.8
All items less food and energy	121.9	127.3	127.6	128.3	129.1	129.6	129.7	130.1	130.8	131.8	132.2	132.3	132.7	133.3	134.1
Commodities less food and energy	114.7	118.6	117.9	119.0	120.5	120.5	120.2	119.9	120.8	120.8	122.0	122.2	121.9	121.7	122.0
Energy commodities	80.9	88.2	90.2	88.4	87.2	87.2	86.4	93.9	91.4	89.8	89.1	92.5	94.1	93.6	102.6
Services less energy	127.0	133.4	134.4	134.8	135.5	136.0	136.4	137.3	137.8	138.8	139.1	139.4	140.3	141.3	142.5
Purchasing power of the consumer dollar:																
1982-84=$1.00	85.5	81.6	81.2	80.9	80.5	80.4	80.3	79.4	79.1	78.7	78.5	78.4	78.0	77.7	77.0
1967=$1.00	28.7	27.4	27.2	27.2	27.0	27.0	26.9	26.7	26.6	26.4	26.4	26.3	26.2	26.1	25.8

Consumer Price Index: U.S. city average and available local area data: all items

(1982-84 = 100, unless otherwise indicated)

Area[1]	Pricing schedule[2]	All Urban Consumers									Urban Wage Earners								
		1989		1990						1989		1990						
		Aug.	Sept.	Apr.	May	June	July	Aug.	Aug.	Sept.	Apr.	May	June	July	Aug.
U.S. city average	M	124.6	125.0	128.9	129.2	129.9	130.4	131.6	123.2	123.6	127.3	127.5	128.3	128.7	129.9
Region and area size[3]															
Northeast urban	M	129.1	130.0	134.5	134.7	134.9	136.0	137.4	128.0	128.8	133.1	133.3	133.6	134.6	135.8
Size A - More than 1,200,000	M	129.5	130.6	135.4	135.4	135.4	136.7	138.0	127.5	128.7	133.1	133.1	133.3	134.3	135.5
Size B - 500,000 to 1,200,000	M	129.1	128.9	133.5	133.6	134.4	135.2	137.2	127.9	127.6	132.0	132.1	132.9	133.8	135.6
Size C - 50,000 to 500,000	M	127.8	128.1	132.0	132.5	133.4	133.9	134.6	130.2	130.8	134.4	134.0	135.7	136.1	130.0
North Central urban	M	122.0	122.5	125.8	126.0	126.9	128.9	128.4	120.0	120.4	123.7	123.0	124.8	124.7	126.3
Size A - More than 1,200,000	M	123.5	124.1	127.3	127.4	128.6	128.6	129.9	120.7	121.2	124.4	124.4	125.6	125.6	127.0
Size B - 360,000 to 1,200,000	M	120.9	121.0	124.8	125.3	125.6	125.8	127.6	118.6	118.6	122.3	122.8	123.1	123.2	125.2
Size C - 50,000 to 360,000	M	122.1	122.2	125.6	125.9	126.5	126.2	127.8	120.8	120.9	124.4	124.6	125.2	124.8	126.5
Size D - Nonmetropolitan (less than 50,0000	M	117.1	117.8	121.1	121.4	122.3	122.6	124.1	116.9	117.7	120.6	121.1	122.0	122.2	123.9
	M	122.1	122.5	126.1	126.5	127.3	127.8	128.7	121.6	121.9	125.3	125.6	126.4	126.9	127.8
South urban															
Size A - More than 1,200,000	M	122.8	123.5	126.8	127.1	127.8	128.6	129.0	122.0	122.5	125.6	125.9	126.7	127.3	127.8
Size B - 450,000 to 1,200,000	M	123.4	123.9	127.4	128.0	128.2	128.6	129.8	121.2	121.7	124.8	125.4	125.7	126.1	127.3
Size C - 50,000 to 450,000	M	121.0	120.9	124.6	124.5	125.3	126.0	127.6	121.6	121.5	125.0	124.9	125.7	126.3	128.0
Size D - Nonmetropolitan (less than 50,000)	M	120.0	120.2	125.3	125.8	128.2	128.0	128.5	121.1	121.0	126.0	126.4	128.5	128.4	129.0
	M	125.3	125.6	129.6	130.0	130.8	131.3	132.2	123.9	124.2	128.0	128.3	129.1	129.6	130.4
West urban															
Size A - More than 1,250,000	M	127.1	127.5	131.5	132.0	132.6	133.1	133.9	124.3	124.6	128.4	128.9	129.4	129.9	130.7

Consumer Price Index

	M	122.6	122.6	126.2	126.4	127.7	128.8	130.0	121.9	122.1	125.5	125.7	126.8	127.8	129.1
Size C - 50,000 to 330,000															
Size classes:															
A (12/86=100)	M	113.2	113.8	117.4	117.5	118.1	118.7	119.6	113.1	113.7	117.1	117.2	117.8	118.3	119.3
B	M	124.0	124.2	128.1	128.5	129.0	129.6	130.8	122.6	122.0	126.4	126.8	127.4	127.8	129.2
C	M	122.9	122.9	126.5	126.7	127.5	128.0	129.4	123.1	123.3	126.7	126.9	127.7	128.0	129.5
D	M	120.5	120.8	125.0	125.6	127.0	127.2	128.2	120.9	121.2	125.2	125.6	126.9	127.1	128.2
Selected local areas															
Chicago, IL-Northwestern IN	M	126.4	127.1	130.4	130.4	131.7	132.0	133.2	122.5	123.1	126.5	126.5	127.9	128.0	129.3
Los Angeles-Long Beach, Anaheim, CA	M	128.9	130.1	134.2	134.6	135.0	135.6	136.3	125.5	126.5	130.2	130.7	131.1	131.6	132.3
New York, NY-Northeastern NJ	M	130.9	132.2	137.3	137.2	137.1	138.4	140.0	128.9	130.3	135.0	134.9	135.0	136.0	137.4
Philadelphia, PA-NJ	M	129.1	130.2	134.3	134.6	135.1	136.3	137.3	129.3	130.4	134.4	134.9	135.5	136.6	137.5
San Francisco-Oakland, CA	M	128.1	126.8	130.7	130.8	131.6	132.3	133.1	127.0	126.1	129.8	129.9	130.7	131.3	132.0
Baltimore, MD	M	—	125.9	—	129.0	—	130.2	—	—	125.4	—	128.3	—	129.5	—
Boston, MA	1	—	132.2	—	137.0	—	138.0	—	—	132.6	—	137.3	—	137.9	—
Cleveland, OH	1	—	123.7	—	128.1	—	128.8	—	—	118.2	—	122.1	—	122.7	—
Miami, FL	1	—	122.9	—	126.4	—	128.7	—	—	121.4	—	124.6	—	126.7	—
St. Louis, MO-IL	1	—	123.9	—	126.7	—	128.0	—	—	123.5	—	126.0	—	127.3	—
Washington, DC-MD-VA	1	—	130.1	—	134.0	—	135.7	—	—	129.5	—	132.8	—	134.6	—
Dallas-Ft. Worth, TX	1	120.0	122.9	—	123.8	—	—	126.0	119.8	—	122.2	—	123.2	—	125.4
Detroit, MI	2	122.2	126.9	—	127.7	—	—	129.4	119.2	—	123.9	—	124.7	—	126.5
Houston, TX	2	114.4	118.3	—	119.7	—	—	121.5	114.9	—	118.6	—	120.0	—	121.9
Pittsburgh, PA	2	120.8	124.9	—	125.0	—	—	127.1	116.0	—	120.1	—	120.3	—	122.0

[1] Area is the Consolidated Metropolitan Statistical Area (CMSA), exclusive of farms and military. Area definitions are those established by the Office of Management and Budget in 1983, except for Boston-Lawrence-Salem, MA-NH Area (excludes Monroe County); and Milwaukee, WI Area (includes only the Milwaukee MSA). Definitions do not include revisions made since 1983.

[2] Foods, fuels, and several other items priced every month in all areas; most other goods and services priced as indicated.

M - Every month.

1 - January, March, May, July, September, and November.

2 - February, April, June, August, October, and December.

[3] Regions are defined as the four Census regions.

— Data not available.

NOTE: Local area CPI indexes are byproducts of the national CPI program. Because each local index is a small subset of the national index, it has a smaller sample size and is, therefore, subject to substantially more sampling and other measurement error than the national index. As a result, local area indexes show greater volatility than the national index, although their long-term trends are quite similar. Therefore, the Bureau of Labor Statistics strongly urges users to consider adopting the national average CPI for use in escalator clauses.

Annual data: Consumer Price Index, U.S. city average, all items and major groups

(1982-84=100)

Series	1981	1982	1983	1984	1985	1986	1987	1988	1989
Consumer Price Index for All Urban Consumers:									
All items:									
Index	90.9	96.5	99.6	103.9	107.6	109.6	113.6	118.3	124.0
Percent change	10.3	6.2	3.2	4.3	3.6	1.9	3.6	4.1	4.8
Food and beverages:									
Index	93.5	97.3	99.5	103.2	105.6	109.1	113.5	118.2	124.9
Percent change	7.8	4.1	2.3	3.7	2.3	3.3	4.0	4.1	5.7
Housing:									
Index	90.4	96.9	99.5	103.6	107.7	110.9	114.2	118.5	123.0
Percent change	11.5	7.2	2.7	4.1	4.0	3.0	3.0	3.8	3.8
Apparel and upkeep:									
Index	95.3	97.8	100.2	102.1	105.0	105.9	110.6	115.4	118.6
Percent change	4.8	2.6	2.5	1.9	2.8	.9	4.4	4.3	2.8
Transportation:									
Index	93.2	97.0	99.3	103.7	106.4	102.3	105.4	108.7	114.1
Percent change	12.2	4.1	2.4	4.4	2.6	-3.9	3.0	3.1	5.0
Medical care:									
Index	82.9	92.5	100.6	106.8	113.5	122.0	130.1	138.6	149.3
Percent change	10.7	11.6	8.8	6.2	6.3	7.5	6.6	6.5	7.7
Entertainment:									
Index	90.1	96.0	100.1	103.8	107.9	111.6	115.3	120.3	126.5
Percent change	7.8	6.5	4.3	3.7	3.9	3.4	3.3	4.3	5.2
Other goods and services:									
Index	82.6	91.1	101.1	107.9	114.5	121.4	128.5	137.0	147.7
Percent change	9.8	10.3	11.0	6.7	6.1	6.0	5.8	6.6	7.8
Consumer Price Index for Urban Wage Earners and Clerical Workers:									
All items:									
Index	91.4	96.9	99.8	103.3	106.9	108.6	112.5	117.0	122.6
Percent change	10.3	6.0	3.0	3.5	3.5	1.6	3.6	4.0	4.8

Appendix 24
Standard Paper Sizes

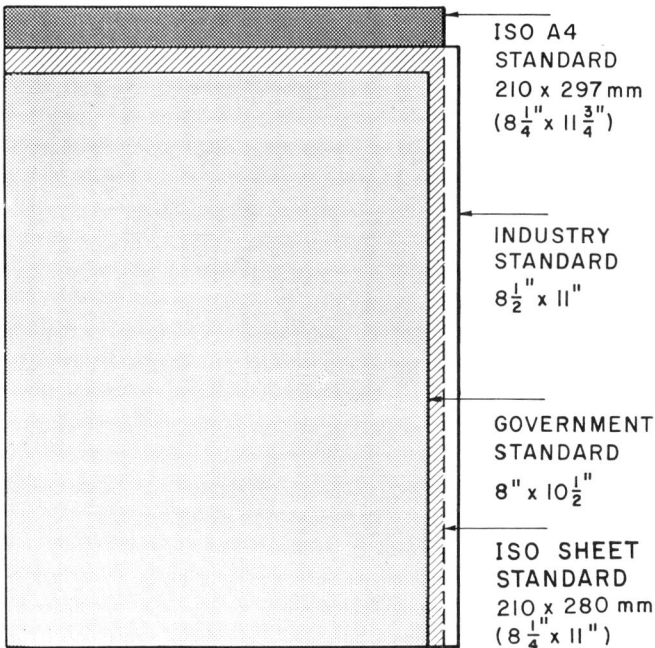

STANDARD PAPER SIZES:

- ISO A4 STANDARD 210 x 297 mm ($8\frac{1}{4}" \times 11\frac{3}{4}"$)
- INDUSTRY STANDARD $8\frac{1}{2}" \times 11"$
- GOVERNMENT STANDARD $8" \times 10\frac{1}{2}"$
- ISO SHEET STANDARD 210 x 280 mm ($8\frac{1}{4}" \times 11"$)

Source: John L. Frirer, *SI Metric Handbook*, New York: Charles Scribner's Sons, 1977, Figure 19-1.

INDEX

A

Abscissa, 44
Accuracy, *see* Confidence interval; Confidence level
Alpha level, 68
Annualizing rates of change
　compounding rates (decreasing), 91
　compounding rates (increasing), 90
　decreasing, 95–96
　defined, 89
　increasing, 92–93
　method, 89–90
Arithmetic average, 3–5
Association, 59

B

Bar charts, 27
Benefit-cost analysis, 192–193
Beta, 68
Bivariate tabular analysis, 16

C

Cartoons, 39
Causality, 45
Chi square, 57–58, 60
　estimating sample size and, 162–166
　formula, 58–59
　interpreting, 67
　method, 58–59
Cluster sampling, 120–123
Coefficient of variation, 8–9
Competitive effect, 101
Complementarities
　net benefit evaluation and, 193
Compounding, principle of, 89, 90
Compounding rates
　decreasing, 91
　increasing, 90
Confidence interval, 133, 134, *see also* Confidence level
　sample size and, 146, 152–155
　standard error and, 145
Confidence level, *see also* Confidence interval
　defined, 133–134
　method, 134–136
　population means and, 144
　proportion above/below values and, 139–141
　proportion between values and, 136–139

Confidence level (*Cont.*)
  p-value and, 146–148
  representativeness and, 145
  sample size and, 146, 152–155
  sampling and, 134–136
  standard error and, 145
Confidence limits, use of, 146
Constant share model, 86
Consumer price index (CPI), 106–107
Continuous data, 27, 28
Correlation analysis, *see also* Correlation coefficients; Gamma
  defined, 49
Correlation coefficients, 49, *see also* Correlation analysis; Gamma
  cautions of, 55–56
  method, 50–51
Cost of living estimates, multiplier analysis and, 106–107
CPI, 106–107
CPI-U, 107
CPI-U X1, 107
Curvilinear (exponential) relationships, 46–47

D

Data accuracy, *see* Confidence interval; Confidence level
Data quality, sampling and, 127
Data set, 3
Davis, James, 52
Degrees of freedom, 60–61
Dependent variable, 14, 44
Descriptive statistics, 9
  defined, 3
Differential shift, 101
Direct (positive) linear relationships, 45
Discounting, 190
Discount rate, 191
Disproportionate sampling, 123–124
Distribution, 3
Dot diagrams, 29
DRASTIC index, 185–186

E

Economic base study, 109–110
Economic multipliers, 105
  multiplier analysis and, 109–111
Equity concerns, net benefit evaluation and, 193
Error level, 68
Estimating
  population means, 144
  representativeness, 145
  sample size for a mean, 152–155
  sample size for mean with large standard deviation, 154–155
  sample size for percentage, 156–160
  sample size for tabular analysis, 160–162
  sample size with chi square, 162–166
Expected cell frequencies (ECF), 38–39
Exponential projection model, 80–83
  modified, 83–85
Externalities, net benefit evaluation and, 193

F

Flow diagrams, 39
Frequency distribution, 6
Frequency polygon, 30
Frequency table, 5, 6

G

Gamma, 49, 50–51, 57, *see also* Correlation coefficients
  cautions, extensions of, 55–56
  quick analysis and, 55
  tabular data and, 50
Graphic techniques
  bar charts, 27
  cartoons, 39
  defined, 25
  dot diagrams, 29
  histograms, 28

Index

Graphic techniques (*Cont.*)
  maps, 26, 36
  method, 25–26
  pie charts, 26
  population pyramids, 34
  process diagrams, 38
  time series diagrams, 30
Grouped data, 4

H

Histograms, 28

I

Incidence of benefits and costs, net benefit evaluation and, 193
Independent variables, 14, 44
Index
  consumer price, 106–107
  defined, 179
  DRASTIC, 185–186
  method, 180–181
  socioeconomic (SES) status, 180
  standardized, formula, 182–184
  weightings and, 181
  z-score and, 180–181
Indices, *see* Index
Inferential statistics, 10–11
Inflation, net benefit evaluation and, 194
Input–output analysis, 109–110
Interval scale data, 10
Inverse (negative) linear relationships, 46

L

Linear projection, 78–80
Linear regression analysis, 47
Line graphs, 30
Location quotient
  defined, 173
  method, 174

M

Maps, 36
Mean, 3–5

Measures of central tendency, 3
Measures of dispersion, 3, 8–9
Measures of significance, 57
Median, 5
Minimum cell size formula, 160
Mix effect, 100
Mode, 5–6
Modified exponential projection model, 83–85, *see also* Exponential projection model
Multiplier, 105
Multiplier analysis
  cost of living estimates and, 106–107
  defined, 105
  economic, 109–111
  standards and, 105–106
Multivariate tabular analysis, 16

N

Net benefit evaluation, *see also* Net present value
  defined, 189
  method, 190–191
  problems of, 193–194
Net present value, 190–191, 192, *see also* Net benefit evaluation
Nominal scale data, 9
Non-continuous data, 27
Nonparametric statistical procedures, 11
Normal curve, 133, 134–135, 142–143
Normal curve table, 135–136
Normal distribution, 10, 142
Normality, 10
Null hypothesis, 62, 67–68

O

Ordinal scale data, 9
Ordinate axis, 44
Outlier, 47

P

Parametric statistical procedures, 10–11
Pie charts, 26
Population sampling, 141–144

Population means, 4
   estimating, 144
Population pyramids, 34
Present value, 190
Process diagrams, 38
Projection techniques
   defined, 75–76
   exponential projection model, 80–83
   linear projection, 78–80
   method, 76–86
   modified exponential projection model, 83–85
   ratio projection model, 85–86
   trend extrapolation projection models, 76–77
Proportionality shift, 100
Published standards, 105–106
p-value, 68–69
   application of, 146–148

Q

Quick analysis, 47, 49
   gamma and, 55

R

Randomness, 116
Random number table, 117
Random sampling, 116, 141–142
Range, 6
Rates of change, *see* Annualizing rates of change
Ratio projection models, 85–86
Ratio scale data, 10
Representativeness, 116
   estimating, 145

S

Sample mean, 4
Sample size
   using chi square and, 162–166
   confidence and, 146
   defined, 151
   determination of, 167–169
   estimating, for mean, 152–155
   estimating, for mean with large standard deviation, 154–155
   estimating, for percentage, 156–160
   method, 151–152
   significance level and, 65
   tabular analysis and, 160–162
Sampling, 145
   cluster, 120–123
   confidence levels and, 134–136
   data quality and, 127
   defined, 115–116
   disproportionate, 123–124
   distribution, 145
   method, 116
   population and, 141–144
   random, 141–142
   simple random, 116–118
   stratified, 119–120
   systematic, 118–119
   weighting and, 124–127
Sawicki, David S., 25
Scales of measurement, 9
Scatter diagrams, 30, *see also* Scatterplots
Scattergrams, *see* Scatterplots
Scatterplots, 14, 29, 30, 44–47
   defined, 43
   method, 43–44
SEM, 152–153
SES, 180
Shift-share analysis
   defined, 99
   method, 99–101
SIC code, 173
Significance, sample size and, 65
Significance level, 68, *see also* Confidence level
Simple random sampling, 116–118
Socioeconomic status (SES) index, 180
Standard deviation, 8, 145
Standard error, 145
Standard error of a percentage formula, 156–160
Standard error of the mean (SEM), 152–153
Standardized index formula, 182–184
Standards
   multiplier analysis and, 105–106

Index

Statistical differences, interpreting, 67–68
Statistical sampling, *see* Sampling
Statistical significance, defined, 57–58, *see also* Chi square
Step-down model, 86
Straight line projection model, 78–80
Stratified sampling, 119–120
Surveys, *see* Sampling
Systematic sampling, 118–119

T

Tabular analysis
  bivariate, 16
  defined, 13
  gamma and, 49, 50
  method, 13–15
  multivariate, 16
  sample size and, 160–162
  table, rules of, 14–15
Time series diagrams, 30
Time-series plot, trend extrapolation projection models and, 76–77
Total shift, formula, 99–100
Trend extrapolation models, 76–85
  exponential projection models and, 80–83
  linear projection and, 78–80
  modified exponential projection models and, 83–85
  time-series plot and, 76–77
Type I error, 68
Type II error, 68

V

Variance, 6–8

W

Weightfactor, 125–127
Weighting, 124–127
  indexes and, 181

Y

Yule's Q, 49

Z

z scores, 135
  indexes and, 180–181